Praise for *Major Label Mastering*

"Absolutely essential for anyone seriously considering any aspect of mastering as their passion or career choice […] this is an under-the-hood approach which anyone involved in mastering can use to understand the entire process."

—Greg Calbi, Sterling Sound

"One of the first songs that Evren mastered for me was a very open, old-school sounding, Ray Charles-esque type of song. He suggested using one of the classic Capitol Records echo chambers on the mix. My first thought was, 'thats insane!', but my second thought was, 'thats cool!, let's do it!' I realized very quickly that I was working with a very creative, open-minded Mastering Guru, who was going to take my mixes to a new level of creativity. Since then, Evren has mastered well over 800 songs for me. In addition to the great sounding final master, what I value about Evren is that he understands the importance of building the relationship between the Mixer and Mastering Engineer. Over the years my mixes have evolved and improved based on what I've learned from Evren and his creative approach, techniques and ideas that he has added to my mixes. It's as if we're on the same football team and I pass him the ball around the 30 yard line so to speak, and Evren runs it in the rest of the way for the goal! Now he has put his techniques and creative approach to Mastering in writing. This book is a must-have for any aspiring student or professional alike seeking a better understanding of the fine art of Mastering!"

—Bill Appleberry, Producer, Mixer (*The Voice*, The Wallflowers, Operator, Puddle of Mudd, Stone Temple Pilots)

T0144530

Major Label Mastering

Major Label Mastering: Professional Mastering Process distills 25 years of mastering experience at Capitol Records into practical understandings and reliable systems. Containing unparalleled insights, this book reveals the mastering tricks and techniques used by Evren Göknar at one of the world's most notable record labels.

Beginning with the requisite competencies every Mastering Engineer must develop, *Major Label Mastering* delves into the particulars of the mastering studio, as well as fundamental mastering tools. Included among these tools is The Five Step Mastering Process, a rigorously tested system that equips the practitioner to successfully and confidently master a project to exacting standards of audio fidelity. Covering all bases, the book discusses both macro and micro considerations: from mindset approach and connecting with clients down to detailed guidelines for processing audio, advanced methods, and audio restoration. Each chapter ends with exercises intended to deepen understanding and skill, or to supplement course study.

Suitable for all levels, this is a unique resource for students, artists, and recording and Mastering Engineers alike. *Major Label Mastering* is supplemented by digital resources including audio examples and video tutorials.

Evren Göknar is a Grammy® Award-winning Mastering Engineer who has been employed at Capitol Studios in Hollywood for 25 years. His work is sought after worldwide, with his mastering credits including legendary recordings and artists ranging from Mariah Carey to KISS, along with hit television shows *The Voice* and *Songland*.

Major Label Mastering

Professional Mastering Process

Evren Göknar

Routledge
Taylor & Francis Group

NEW YORK AND LONDON

First published 2020
by Routledge
52 Vanderbilt Avenue, New York, NY 10017

and by Routledge
2 Park Square, Milton Park, Abingdon, Oxon, OX14 4RN

Routledge is an imprint of the Taylor & Francis Group, an informa business

Library of Congress Cataloging-in-Publication Data
Names: Göknar, Evren, author.
Title: Major label mastering : professional mastering process / Evren Göknar.
Description: New York : Routledge, 2020. | Includes index.
Identifiers: LCCN 2019043697 (print) | LCCN 2019043698 (ebook) |
    ISBN 9781138058576 (hardback) | ISBN 9781138058583 (paperback) |
    ISBN 9781315164106 (ebook) | ISBN 9781351673143 (adobe pdf) |
    ISBN 9781351673129 (mobi) | ISBN 9781351673136 (epub)
Subjects: LCSH: Mastering (Sound recordings) | Sound recordings—Production
    and direction.
Classification: LCC ML3790 .G644 2020 (print) | LCC ML3790 (ebook) | DDC
    621.389/3—dc23
LC record available at https://lccn.loc.gov/2019043697
LC ebook record available at https://lccn.loc.gov/2019043698

ISBN: 978-1-138-05857-6 (hbk)
ISBN: 978-1-138-05858-3 (pbk)
ISBN: 978-1-315-16410-6 (ebk)

Typeset in ITC Giovanni
by Apex CoVantage, LLC

Visit the eResources: www.routledge.com/9781138058583

For Esma and Milas

Contents

System is that principle of order by which confusion is rendered impossible.*

—James Allen, *Eight Pillars of Prosperity*

* James Allen (1911) *Eight Pillars of Prosperity: Fourth Pillar—System*, New York, NY: Penguin. p. 80. Penguin Edition 2009.

Preface

I'VE JUST SEEN A PLACE

The first time I saw the Capitol Records building in person was in December of 1983. I was 17 years old, a senior in high school and had traveled west from my home in Detroit, Michigan to visit my sister in San Francisco for the Christmas holiday. We drove to Los Angeles to explore Hollywood. I still remember driving south on the 101 freeway and passing the iconic round tower. I became immediately transfixed by its unmistakable atomic-age architecture—it appeared to float above a rectangular base like a Cold War-era rocket ready for launch. Capped with a needle-like spire and those massive letters (in case amnesia struck?) . . . it was the definition of a head-turner.

I recalled numerous record albums, including my favorites by The Beatles, Pink Floyd, Bob Seger, and others, all with the familiar oval logo printed on them. The rock radio stations in Detroit continually played recordings from artists on the Capitol Records label. I wondered if and when members of those bands had been there, and who might be there as we drove by. I wondered what the place was like inside. It possessed the mystical allure of a holy site, a mid-century curiosity on every Hollywood tour. We continued on with our sightseeing, but I couldn't know then that in 11 years, that building would become my second home, from where I would launch a career as a Mastering Engineer. For that matter, at this early juncture, I didn't even know that mastering existed.

A LONG AND WINDING ROAD . . . TO MASTERING ENGINEER

After college, I moved to Los Angeles to write songs and play in rock bands. In so doing, I implemented the pragmatic idea of seeking work in recording studios, to expand my musical community and access inexpensive studio time for my bands. Over time, this led to learning recording engineering. By the mid-1990s, I had worked as a recording engineer for over five years—primarily at Paramount Recording Studios in Hollywood, but also with producer Rudy Guess and independently at various Los Angeles studios—when I began to develop an interest in audio mastering. The first mastering session I attended was in 1992 with Bernie Grundman on the Carole King album "Colour of Your Dreams" (produced by Guess) that I had assistant engineered. Through research into the prospect of mastering, many résumés mailed to mastering studios and networking, I was ultimately hired by the mastering department at Capitol Records.

INITIAL DUTIES

After working as a recording engineer, my initial duties at Capitol Records seemed mundane, but later as a Mastering Engineer, I understood their incredible value. My job was to duplicate, quality control (QC), and assemble or otherwise prepare compact disc (CD) masters that were in the pulse code modulation (PCM) 1630 ¾" video cassette tape format. For editing, assembly, and QC, I used the Sony DMR-4000 VTR and the Sony DAE-3000 Editor. In addition to QCing, this involved such glamorous tasks as: PQ encoding (entering the start and stop indexes that end up in the P and Q subcode of a compact disc) and generating a PQ sheet, inserting digital black between songs, printing out a clean bit stream analysis (no mutes) of the audio, and labeling the 1630 tape . . . with a typewriter. These projects were generally reissues from the catalog department or mastered by established Mastering Engineers.

I quickly began reframing my recording and mixing skills and applying them into a mastering context. Soon I was performing assemblies from analog tape to digital 1630, including level adjustments and equalization (EQ). This also meant earning the confidence of catalog producers who would later enlist me for remastering seminal albums or new collections from the Capitol Records catalog.

QUALITY CONTROL (QC)

QCing refers to listening down to an entire album or mastering project once it has been completed to verify its integrity (from first to last song). I did this exclusively for about a year while beginning to EQ and master client source tapes. QC remains an essential skill for a Mastering Engineer that involves identifying and articulating artifacts, anomalies, or problems on an audio master. The care and attention to detail honed in QC translates to other critical mastering skills.

THE FLAT TRANSFER

In order to preserve tapes or create a backup for the library, I would often make a *flat transfer* to a digital or analog format. High-quality transfers pertain to mastering, and represent a fundamental aspect of the art. You play back the audio source and record or capture to create a new one. Again, this could be to make a production master or a safety of the original—especially commonplace if the original is deteriorating or damaged.

THE ASSEMBLY

Often, the various departments at Capitol from front line to catalog would require masters to be assembled for new releases. Previously mastered audio from various albums would be compiled into a new master, known as *assembling* a master. With an *assembly*, leveling of the album to a target level is required, and occasionally some light equalization (EQ), but generally it means compiling the tracks in sequence for a new master. Practicing this by loading a set of songs into your digital audio workstation (DAW), then matching their volumes to a selected target level and generating a compilation, represents an excellent exercise. As a component of mastering, compiling *assemblies* is a good way to understand additional enduring mastering concepts such as song-to-song cohesion in both frequency response and level.

SUMMARY OF SUBSIDIARY DISCIPLINES

Command of these three subsidiary disciplines of mastering proved essential to my evolution in professional audio. The hierarchy of tasks and job classifications at

Capitol provided fertile ground for development into the role of Mastering Engineer. Compounded with my recording experience, I understood how the patience and skills developed with these tasks would inform my habits as a Mastering Engineer. One of the issues with the proliferation of technology, and more people embarking on audio careers, remains that they tend to skip time spent developing foundational skills. I admonish neophyte engineers to take heed—develop skills in QC, transferring between formats, and the assembly of compilations or CD Masters.

PAPERBACK WRITER . . . ?

I came to write this book after initially writing a comprehensive outline for a ten-session mastering workshop. It began by answering the question, "What is mastering?" and continued through the various steps of what actually takes place in a mastering session. Subsequently, I gave a one-hour talk on mastering for the Audio Engineering Society club at California State Polytechnic University Pomona, and a student inquired about the workshop. At the same talk, I met the head of the Music Technology Department there, Arthur Winer, who asked me if I would be amenable to teaching their mastering class. I agreed, and blended my workshop lessons with the existing course syllabus for a robust class on audio mastering.

I found teaching rewarding, and realized that I remained uniquely positioned to impart valuable information to enthusiasts of audio mastering. I was ultimately inspired to write a book on the subject when it became clear that my unique experience mastering at Capitol Records should be documented and available for reference; and that there existed a dearth of published material outlining a comprehensive step-by-step approach to mastering. So, I began to write and committed to completing this technical guide and practical manual about audio mastering.

Evren Göknar
Los Angeles
2019

Acknowledgments

I am indebted to and appreciative of the many talented, hard-working colleagues and studio management staff who glimpsed potential and initiative and supported opportunities for me to work, learn, develop, and thrive for 30 years in professional audio. An exalted category of gratitude is reserved for the artists and musicians who have trusted me with their most precious and deeply prized possessions—their music and recordings. By working with them to understand and help achieve their artistic vision, I have been enriched as a Mastering Engineer and trust that my work has exceeded their expectations for the final mastered result of their recordings.

Ultimately, this book would not exist without the many associates and colleagues over the years who contributed to my professional audio career and daily mastering work.

RUDE STUDE RECORDING (1990–1993)

Rudy Guess (ad astra) and Lorna Guess both for their welcoming encouragement, Rudy for mentoring the 23-year-old Evren, and outlining basic methods for successfully running a recording session from basics to overdubs to mixdown, and for enlisting me to co-engineer the recording of Carole King's "Colour of Your Dreams" album (1993).

PARAMOUNT RECORDING (1989–1995)

Adam Beilenson and Michael Kerns for successfully keeping their studio facility humming through the Los Angeles riots (1992), the Northridge earthquake (1994), and the home studio revolution. But most of all for allowing me to sneak into the studios from my post in the office and learn recording from the existing engineering staff aka *the cats*! Michael Schlesinger (ad astra), Barry Conley, Jamie Seyberth, Keith Barrows, Stoker, Geza X, Voytek Kochanek, Michael Melnick, Michael Becker, Anne Catalino, and Lou Hernandez for being brothers/sister in arms and sharing recording approaches in the days of 2-inch analog tape while we never backed down from sessions with a midnight downbeat!

CAPITOL STUDIOS AND MASTERING (1995–PRESENT)
Executive/Management/Office/Administrative

Michael Frondelli for hiring me and supporting my transition to mastering, Robin Bechtel for a character reference to M.F. Pete Papageorges for allowing me drop off my résumé and supporting my transition to mastering. For friendship and years of session booking and administrative support: Cassandra Spunbarg, Beatrice Olsen, Rob Christie, Cathleen Weinrich, Talin Titizian, and Jordan Koppelman. Greg Parkin and Matt Graber for embracing independent artists/producers as important clients and standing by the engineering staff. Pat Kraus, Kevin Reeves, Roey Hershkovitz, Ryan Simpson, and Paula Salvatore for keeping Capitol Studios and Mastering running like a Swiss watch and managing the legendary facility with care and consideration.

Colleagues/Engineers/Technical Staff

The following Whiz Kids who kept or keep all systems functioning: Jeff Minnich, Tom Ketterer, Bruce Maddocks, Tom Schlum, Denny Thomas, James Goforth (ad astra), Dave Clark, Assen Stoyanov, Peter Gonzales, and Bryant Lewis. Colleagues past and present who shared/share in running recording sessions or completing the label's workflow: Scott Lechner, Jeff Rach, Odea Murphy, Kevin Hayunga, Jay Rannellucci (ad astra), Christina Paakari, Leslie Ann Jones, Chandler Harrod, Steve Genewick, Nick Rives, Jeff Fitzpatrick, Charles Cosin, and Mike Jones. David McEowen and Perry Cunningham, known respectfully as 'the palominos' for their status executing asset preservation, for meticulously archiving and cataloging priceless analog tapes in high-resolution digital.

Capitol Catalog/Independent Catalog Producers

I've worked regularly with these past and present producers who mine the Capitol and affiliated label catalogs for under-represented gems for remastering and re-release from the archive: Greg Ogorzelec, Cheryl Pawelski, Frank Collura, Matt D'Amico, Michael Murphy, Christian Johnson and Kevin Flaherty. Marty Wekser and I worked tirelessly on the Varese Sarabande catalog, and David Tedds and I went back to original analog source tapes for the Grand Funk Railroad, Queensrÿche, and Pat Benatar catalogs.

Mastering

The Capitol inner sanctum of mastering past and present: Kevin Reeves, Wally Traugott (ad astra), Bob Norberg, Ron McMaster, Kevin Bartley, Robert Vosgien, Ian Sefchick, and Mark Chalecki. With a bullpen like this, there's bound to be a friendly skirmish here and there; fortunately, mutual respect prevails at Capitol.

CALIFORNIA STATE POLYTECHNIC UNIVERSITY POMONA (2014–2015)

I must acknowledge Professor Arthur Winer, head of Audio Technology in the Music Department, and Dr. Iris Levine Chair of the Music Department for enlisting me to teach the mastering class at CSPUP. This experience kindled my passion for teaching and inspired me to write this book.

INTERNATIONAL BROTHERHOOD OF ELECTRICAL WORKERS (IBEW) LOCAL 45

Capitol Studios is one of the rare recording facilities whose studio employees are under a collective bargaining agreement. A unique harmony pervades and Elaine Ocasio, Victor

Marrera, Lupe Perez, and formerly Rick Rogers consistently find the common ground so that the audio engineers can continue to focus on making exemplary recordings.

ROUTLEDGE/TAYLOR & FRANCIS GROUP/ FOCAL PRESS

I'm grateful to the detailed work and focus from my publisher including Hannah Rowe, Shannon Neill, and beginning with Lara Zoble. The book benefited from their high standards and patience in receiving the manuscript.

MY FAMILY

My remarkable and loving family: brilliant wife Dr. Luna Merve Göknar, force-to-be-reckoned-with son Milas Göknar, and effervescent-sunbeam daughter Esma Göknar. Throughout this entire intense writing process, they never wavered in support.

MY PARENTS

I credit any steadfastness and fortitude of my own to my parents Dr. Meral Ülker Îli Göknar and Dr. Mehmet Kemal Göknar. Their example of perseverance was relied on continually to complete *Major Label Mastering: Professional Mastering Process*.

Credits/Gratitude/Disclaimer

I would like to thank the following individuals for their specialized contributions to this book:

> Citations, motivation, and writing approaches: Dr. Erdağ Göknar (my twin brother)
> Initial copy editing of the entire manuscript and all block diagrams: Mark Brookner
> Photography and editing: Aydın T. Palabıyıkoğlu
> Additional photography: Evren Göknar
> Graphic design (including the book cover, Figure 15.1 creation, and additional photograph editing): Jill K. Avenaim

AUTHOR DISCLAIMER

All figures, images, photos, or mentions of equipment have been personally vetted by me and used by permission. I was not remunerated by any company, manufacturer, or engineer whose products appear in this book.

Introduction

MASTERING DEFINED

Mastering is the process of optimizing audio to create the definitive final version of a recording. It represents the critical last step in the recording process before replication at a plant or distribution to the consumer. A master is used to create compact discs, vinyl cutting masters, and digital files for streaming or download. Additionally, mastering represents a discipline that requires a unique intersection of both creative and technical skills. An excellent Mastering Engineer possesses a rare intersection of musical knowledge, technical acumen, and recording expertise. They are sought after and relied upon to tastefully enhance—or judiciously leave alone—final mixes. Ideally, the result will present the recording in the best possible audio fidelity.

MAJOR LABEL MASTERING: PROFESSIONAL MASTERING PROCESS

This book is organized into five main parts. Part I explores The Key Competencies and Understandings one must possess, learn or develop in order to effectively master audio. This section presents a checklist for a Mastering Engineer (aspiring or established) to assess their particular strengths or weaknesses with, and to work toward continually evolving and providing high fidelity results successfully. Here, the critical concept of listening experience as detailed by The Eleven Qualities of Superb Audio Fidelity is thoroughly investigated. Part II elucidates The Mastering Studio and Fundamental Mastering Tools. In this section I introduce the concept of The Three Zone Mastering System. Part III delves into The Five Step Mastering Process that methodically takes the source file from flat mix to completed master. Beginning with the substructure of objective and subjective assessment of the source audio, and implementing the game plan for processing audio via gain structure, equalization, compression and limiting— the Mastering Engineer achieves the desired result and impact at the genre-appropriate target level. Part IV investigates macro considerations that inform a mastering session including: steps before beginning the project, mindset approaches to mastering, and helpful mastering secrets. Finally, Part V explores micro considerations of the mastering process in more detail by building on the basic tools and methods, and introducing advanced tools and methods used in professional mastering. Note that many chapters conclude with exercises to help verify understanding of key concepts. I've added these so the book can be used in a course/classroom context or for self-study.

Major Label Mastering distills my 25 years of professional mastering experience in the discipline into practical and repeatable approaches. First, a mastering student or enthusiast must endeavor to understand the mastering process and set up a professional full-range listening environment. With consistent practice, a baseline of expertise with recordings and mastering tools will be achieved. As understanding evolves, they can avail themselves of my time-tested Major Label Mastering Process encapsulated in the following Five Steps from Part III: 1) objective assessment of the source file, 2) subjective assessment of the source file, 3) The Mastering Game Plan (tool selection/ processing), 4) assembly, and 5) delivery of requested audio formats. If a student of mastering follows this formula regularly, and supplements it with authentic musicality and discerning ears, then they will achieve mastering excellence.

The title includes 'major label' because that is the nexus of my professional mastering experience, and there are some unique contextual aspects therein that I trust remain beneficial to all Mastering Engineers. That fact notwithstanding, the information presented here is relevant for any aspiring or practicing Mastering Engineer to enhance their knowledge of and focus on the discipline. The information applies to anyone from a music technology instructor, audio enthusiast, audiophile, independent artist, engineer, producer, or label, to an established Mastering Engineer seeking to compare approaches, try new concepts, or simply expand their audio mastering acumen. I advise the reader to initially read the book parts and chapters in sequence, as the information presented earlier sets the foundation for what follows. Subsequent to the first reading, the sections are designed to be useful as a reference. In that regard, Appendix I is a comprehensive list of common mastering terms and acronyms. Indeed, a central motivation of mine is to demystify and democratize the cryptic art of audio mastering!

PART I

The Key Competencies and Understandings of a Mastering Engineer

A Mastering Engineer must endeavor to develop critical audio analysis and processing skills, and also understand fundamental principles of digital and analog recording. Once a baseline of proficiency is achieved in these areas, the conceptualization and execution of audio mastering becomes readily available, and the audio fidelity of completed mastering projects drastically improves.

CHAPTER 1

The Mastering Engineer

COMPETENCIES

In this chapter, I will identify and explore The Ten Competencies of a Mastering Engineer. This represents a bird's-eye view for those seeking a greater understanding of or considering a career in audio mastering, or a checklist for established or practicing Mastering Engineers. Broadly speaking, audio mastering represents a discipline that requires sophisticated listening abilities, familiarity with musical genres, technological acumen, and an awareness of musicianship. More specifically, a dedicated and effective Mastering Engineer must understand, practice, and execute a set of component skills that constitute the greater discipline of audio mastering. The array of these skills is expansive but can be organized into the following *ten competencies*, one must become:

1. An Audiophile
2. A Critical Listener
3. Aware of Recording Methods and Technologies
4. Knowledgeable About Audio Mastering Tools
5. Informed About Acoustics and Listening Environments
6. Proficient With Digital Audio Workstations
7. Adept at Quality Control (QC) Methods
8. Able to Generate the Audio Master
9. Implement Recallable Workflow Approaches
10. Personable and Entrepreneurial in Demeanor

This represents a comprehensive list and no aspiring Mastering Engineer begins with proficiency in each area. The process of regular engagement with audio, mastering, and seeking expertise such as a community of Mastering Engineers, or this and other pertinent books from reliable sources, will activate and develop them. Some of the listed skills may come more naturally to you than others, but fortunately all of them can be learned or improved upon, and ultimately achieved with time and dedication. Engaging in and developing these skills will allow you to offer the client insight into audio fidelity enhancement and prepare you to interact effectively with them. They must feel confident that you understand their sonic vision for the finished product, so you must be an effective listener to their requests or concerns. As the critical last stop in music production, mastering involves great responsibilities and the Mastering Engineer remains accountable to the artist, engineer, producer, and record label in completing each project effectively.

Once you are mastering, you may receive fantastic feedback, and feel great about your results, but you will be best served by a mindset of providing excellent audio fidelity while continually evolving. Allow your network of clients, the quality of your work and studio/facility, and your accolades to represent your success. Confidence in your abilities accompanied by a beginner's mind (*shoshin*[1] in Zen Buddhism)—engaging the totality of your sensibilities—represents an effective mindset for a Mastering Engineer. This way, you remain open to new approaches and ideas as you work and evolve. Whereas your opinion will clearly matter, understanding an artist's musical vision and delivering a great sounding, high fidelity master will allow you to meet or exceed their expectations and result in referrals, repeat customers, and an excellent reputation. Let's now explore the ten competencies in greater detail.

Become an Audiophile

Embrace becoming an *audiophile*, literally "audio lover" in Latin—*audire* means to hear, and *phile* means to love. In doing so, you will increase your awareness and understanding of the quality of recordings, the common formats and resolution options for digital audio, and the quality sound playback systems that reproduce them. This concept underlies and informs the ethos of audio mastering. Audiophile values include a pure connection with and sensitivity to the music, ideally reproduced on a transparent playback system with minimum hardware component coloration, revealing only the music captured in the recording. By engaging in some basic research, you will become knowledgeable about both analog and digital audio playback devices and components (cables/interconnects, software players of digital files, turntables, computer input/output [I/O] cards, digital-to-analog [DA] converters, audio cables, power amplifiers, pre-amps/receivers, speakers/drivers, speaker crossovers, and electronic components/equipment design).[2] If you are new to this concept, any issue of *Stereophile*, *Audiophile*, or *The Absolute Sound* or their corresponding web sites offer both scientific and subjective equipment reviews. Although the language and descriptions can get flowery, possessing the correct vernacular remains helpful when communicating likes and dislikes about audio equipment, recordings, and sound. Additionally, it is important to possess an understanding of common *lossy* (data reduced) and *lossless* (data preserved)[3] digital audio formats (*lossy*: .mp3, .aac, and .m4a; *lossless* and *high-definition lossless*: .wav, .aiff, .flac, .alac, and .dsd—used for Super Audio Compact Disc [SACD]).

In a recording and mastering context, the ideology behind audiophile recordings is also relevant. These are often live-to-two-track stereo recordings (utilizing either X-Y, binaural, or Blumlein[4] microphone techniques) done in a highly regarded, acoustically correct space (studio, hall, or venue). Producers of these recordings embrace an ethos of using minimal devices in the recording chain outside of high-quality microphones and microphone preamplifiers, ideally utilizing no compression or equalization, so as to capture the purity of the sound and musical performance with minimal processing and electronics noise. If the music genre doesn't support this purist approach, many music labels have been involved in having popular titles in their catalog remastered from original analog source tapes into high-resolution audio formats: .flac, .alac, .dsd or high-resolution audio (24bit with a sampling frequency at or above the *Red Book* CD standard of 44.1kHz—16bit, usually 48kHz, 88.2kHz, 96kHz, 176.4kHz, or 192kHz).

An example of two record companies that produce and release audiophile recordings are Chesky Records (started by brothers Norman and David Chesky, who also operate a high-resolution audio download platform HDtracks [hdtracks.com]); and Sheffield Labs (started by renowned Mastering Engineer Doug Sax and his brother Sherwood). Both of these labels produced and sell an audiophile test disc (The Sheffield/A2TB Test Disc [SL10508] and the Chesky Jazz Sampler and Audiophile Test CD [JD37]) to assist in developing listening skills and also to test speaker placement and room acoustics for audio image assessment in a listening room. There are many other labels dedicated to high-resolution/audiophile quality recordings (for example, Mobile Fidelity Sound Lab and the Columbia Master-Sound series specialize in half-speed vinyl releases), and research on the HDtracks website provides a good survey and understanding of the high-definition digital audio market. Become familiar with a few of these types of recordings, and keep them in a folder on your digital audio workstation (DAW) to compare against similar projects you're working on.

A Note on Sampling Frequency (S) and Bit Depth (BD) in High-Resolution Audio

I will contextualize the high-resolution audio discussion by framing high *sampling frequency* audio with some objective scientific ideas. Note that the *sampling frequency* (S) determines the *Nyquist frequency* (N) or bandwidth of the audio using the relationship $N=S/2$.[5] This means that digital audio sampled at: 44.1kHz produces a frequency bandwidth of 0Hz–22.05kHz; 48kHz audio, 0Hz–24kHz; 88.2kHz audio, 0Hz–44.1kHz, and so on. Remember that the threshold of human hearing is around 20kHz. Therefore, although the high-resolution file is reproducing more audio, and more audio beyond the audible spectrum, it is not audible to humans. Many people claim to notice a difference in the quality and impact of high-resolution audio, hence the demand and the market. Ideally, you can conduct your own tests comparing digital audio formats and resolutions, and derive independent conclusions.

Bit depth determines dynamic range in decibels (dB) of sampled digital audio at a resolution of 6dB per bit; 16bit digital audio has a dynamic range of 96dB, and 24bit digital audio has a dynamic range of 144dB. There is also a difference between 16bit and 24bit in other subjective audible qualities, namely depth, dimension, and richness of the audio image; 24bit represents the limit of DA conversion, so although many other bit depths are used in host DAW internal processing or storage (32bit and 64bit floating-point are common), the audio remains at 24bit resolution.

Musical Genres

With time spent listening to and working on mastering music, your understanding of and feel for musical genres and styles will develop. Rock, pop, country, singer/songwriter, hip-hop/rap, electronic dance music (EDM), classical, jazz, and their various sub-genres all have specific production characteristics. Basic examples are: pop is vocal forward, rock is often guitar forward, rap is kick/bass forward, and jazz preserves dynamic range. Familiarity with these genres will serve you in making *subjective assessments* (see Chapter 6—Step II: Subjective Assessment) of the source audio, and will also allow you to assess the strengths and weaknesses of the recording. Occasionally, this may result in revision notes for the mix engineer. This insight is important so that you ensure your masters are providing a genre-appropriate *listening experience.*

Musicality

Musicality encompasses other important *audiophile* qualities. Drawing on the sensibilities that being a musician provides, possessing an understanding of song structure and macro-dynamics (meaning a developed awareness of song sections, cadences, passages, themes, dynamics, and musical payoffs) is very advantageous for a successful Mastering Engineer. Once you are manipulating equalizers, compressors, limiters, and other audio processing equipment, a *sensitivity to the music* remains paramount. What is the mood? How is the song recorded? How will your chosen adjustments affect the *listening experience*? Your interest in mastering indicates that you are likely engaging in audiophile activities already, and *deliberate listening* to live or recorded music will further hone your abilities of auditory perception.

Deliberate Listening and Subjective Assessment

In her excellent book *Grit*, Angela Duckworth explores the concept of Deliberate Practice utilized by world-class musicians and athletes over time to become expert at executing either musical prowess or athletic skill. She explains that elite practitioners of a discipline practice with a specific intention.[6] Similarly, the primary concern of a Mastering Engineer when engaged in *deliberate listening* remains, *"is the listening experience pleasing, effective, and genre-appropriate?"* This question, and the answer(s) it conjures in the ears and mind of the Mastering Engineer, remains mission critical to their success in mastering the recording. It is the foundation of *subjective assessment*, Step II in my breakdown of The Five Step Mastering Process that bridges the approach from source file to master (see Chapter 6—Step II: Subjective Assessment).

Your sensibilities, experiments, and experiences as an audiophile will inform your decisions in the mastering studio. They cannot be underestimated in informing the quality of an effective Mastering Engineer.

Become a Critical Listener

Deliberate listening will allow you to assess with confidence the qualities of great audio fidelity. What comprises a sublime listening experience? When does a recording present these desirable characteristics? Understanding this viscerally is a natural response to music and sound. Articulating the elements of that experience—and conversely, what may be lacking from it—is *critical listening*. Our ears receive and our brains (the ultimate audio summing mechanism) process sound, ironically revealing that the recordings we labor over are judged by an entirely subjective listening process. But remember, if your subjective musical sensibilities consistently translate positively to other music creators, you will be in demand as a Mastering Engineer.

By virtue of working on many projects from the Capitol catalog, I am able to hear and work on excellent sounding master tapes of famous recordings that were from the best studios, producers, and engineers of a specific era. Over the years, these projects have informed my *critical listening* and audio analysis skills. Spend time listening to highly regarded audio from a fidelity standpoint to bolster your listening acuity.

The Image or Soundstage

Once a critical listening environment is set up (a listening room, mastering studio, or control room), music will then be presented as a soundstage between the two speakers commonly referred to as the *image*. The *image* consists of a 'solar system of sound' wherein the individual musical elements are defined, discernable, and clear in space, taking advantage of the breadth, height, and depth of the soundstage spatially and frequency-wise. This *spatial dimension* and *depth* are desirable in a great audio recording, and a good Mastering Engineer will preserve or enhance this aspect of the recording. The song sections must ebb and flow or blossom and retreat as the intensity, intention, and emotion of the work dictates. The instruments must be well-balanced, interlocking by each occupying a different placement in the *image* and different or harmoniously co-existing frequency ranges. In the instance of recordings with a vocal, these instruments must generally support a present, natural, full-range, and pleasing vocal tonality. The frequencies also must be well-balanced, and can be visualized with the lower frequencies such as a kick drum near the bottom, moving up higher to bass, guitars, and keyboards through the middle, the vocals rising out of the middle, and cymbals or percussion at the very top, thus giving height to the *image*.

This hierarchy is reinforced by the fact that speaker cabinets are often designed with the woofer on the bottom, and mid-range and high-frequency drivers on top. Although bass frequencies are considered non-directional, as you work on them, they always appear to be establishing a foundation. At your listening position or sweet spot, the frequencies should be time-aligned (arriving at your ears simultaneously), but bear in mind that all frequencies travel at different speeds: lows are slower with longer wavelengths, and highs travel faster and have shorter wavelengths. Additionally, subwoofers, from which very low frequencies emanate, sit on the floor. And finally, in a live performance, the kick drum of a drum set is on the floor, but the cymbals are a bit higher on stands, and the PA system amplifies the vocal from speakers mounted on stands or even the ceiling, giving it height placement.

Playback System Considerations

Richard Harley writes in *The Complete Guide to High-End Audio* that, "each component in a playback (monitoring) system is like a pane of glass, and the more transparent and uncolored, the better".[7] A clean monitoring path for digital file playback consists of the following devices (Figure 1.1).

Please note that as simple as this setup seems, one can easily spend into the many tens of thousands of dollars and beyond for a full-range system suitable for *critical listening*. This same audio monitor path approach is used in a mastering

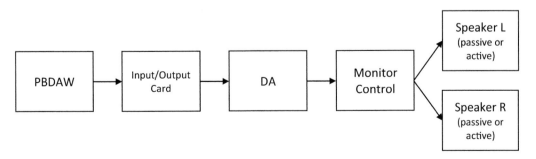

Figure 1.1 Basic playback components chain from DAW to speakers.

studio setup, as well, so it is advantageous to familiarize yourself with and research audiophile playback systems in addition to professional audio setups. Some dealers will have decent listening rooms where you can bring your own audio and audition different speakers or amplifiers. Bear in mind that marketing and common sense don't always align. What is necessary versus what is available often represents vastly different cost structures. As such, be exceedingly demanding but reasonable in your selections.

Researching and owning a great pair of audiophile headphones will also serve you well. Headphones provide an alternate reference option for mastering, and also for the *Quality Control* (QC) of completed masters before sending them to the client, distributor, or replicating facility. Some common options to consider are: open-back (offers a more expansive, natural sound) versus closed-back (offers greater isolation and sound remains immediate), driver response or 'quickness' in reproducing transients in music, frequency response, coloration (headphones, like standalone speakers, can unnaturally color audio), comfort and cost.

If you are reproducing/listening to vinyl albums, then the path is even simpler (Figure 1.2).

Vinyl albums have seen a strong resurgence in recent years. Many aspects of vinyl records contribute to a romantic 'bygone era' appeal. Surface noises and flipping the album over notwithstanding, vinyl albums represent a completely analog playback technology that reproduces true analog audio versus the waveform sampling that occurs in digital audio. Even the frequency response or level limitations of the medium contribute to its listening appeal. Audio that was originally released on vinyl generally remains less compressed and retains dynamics and transients that can get limited out of more modern mastering treatments for digital distribution. Additionally, louder volume levels on a vinyl album result in wider grooves which limit the amount of audio that will fit on an album side. As such, vinyl albums are *generally* around 8 dB quieter in level than modern compact discs. This is a substantial difference in level—if 6dB represents a doubling of amplitude, then a CD can be *perceived* as 100% *louder* than its vinyl counterpart through the same playback system. *Audio level* is measured objectively, but *perceived loudness* is in the subjective realm of *psychoacoustics*,[8] which affects *listening experience*. For instance, mid-range frequencies become more apparent to our ears as profiled in the Fletcher–Munson Equal Loudness Contours (refined by Robinson and Dadson).[9] Further, the mechanics of

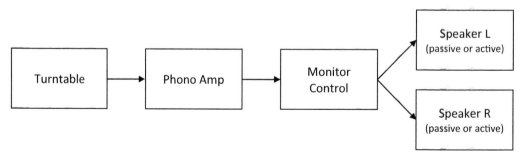

Figure 1.2 Components chain for vinyl album playback. The path is all analog, and does not require the DA converter shown in Figure 1.1.

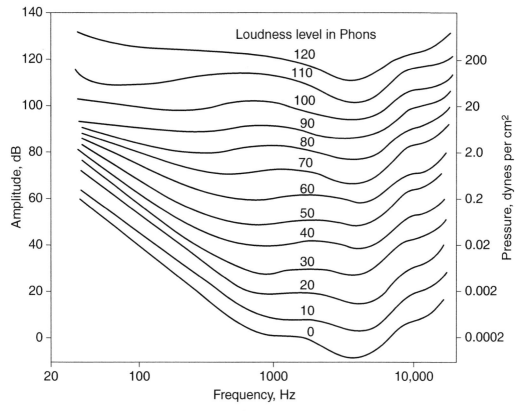

Figure 1.3 Fletcher and Munson showed that perceived loudness (in phons) varies with frequency and sound pressure level (SPL). The curves indicate human hearing is most sensitive between 2kHz–5kHz, the resonant frequency of the ear canal. The ear's frequency response is more uniform at higher loudness levels as shown by flattened curves above 90 phons.

a turntable require limits in the low and high frequencies to maintain groove width and prevent the needle from jumping, often resulting in a warmer sound for vinyl albums, which many music connoisseurs prefer (Figure 1.3).

Become Aware of Recording Methods and Technologies

I referred to the spectrum of recording approaches between a live-to-two-track stereo recording and a modern multi-track recording under item 1—Become an Audiophile earlier. Genre-specific audiophile recordings (often classical, jazz, solo instruments, chorale, singer/songwriter) are recorded live with minimalist stereo microphone techniques, and the engineer deliberately selects the room or hall for its acoustic properties. Adding close-in microphones may further enhance mixdown options, but careless placement can create phase anomalies and undermine the purist approach of an audiophile recording. Each performer's distance from the stereo microphones determines which instrument or element is featured, and which instruments are supporting. The addition of more microphones and their phase relationships, along with the cumulative noise from additional recorded signal, all combine to diminish a purist's stereo recording approach.

Recordings of other genres (rock and hip-hop and their sub-genres, pop, dance, EDM) are multi-tracked and/or use sampled sounds, layered/stacked elements, and created sounds with additional studio effects added at the mix stage of the recording. With consistent practice, you'll be able to discern how a recording is made by the genre, impact, dynamics, and fidelity of the recording. An enduring music production principle is that performance-oriented ensemble recordings, or 'takes' of individual instruments, possess vitality and a natural quality making them more listenable over time than excessively edited or processed recordings. Bear this in mind if you also engineer recording or mixing sessions. A performance requires accomplished musicianship, and technology inadvertently undermines musicianship with temptations such as auto-tuning, take comping, sound design, and midi recording. One can readily produce recordings that exceed the artist's performance ability. There remains an aspect of human performance and its nuances that appeals to our ears. What's important to remember is that if a recording or mix is unpleasant to listen to flat (un-mastered), you will encounter difficulty making it more listenable after mastering. If possible, these observations can be imparted to the artist, engineer, or producer, and adjustments can be requested, yielding a better sounding final master.

Great recordings represent a document of sound to analyze for deeper understanding of a compelling listening experience. Time spent recording and mixing, or even assisting in a professional recording environment, translates into a deeper understanding of the component parts of a finished recording. Eventually, you will be able to impart priceless information on how to improve a recording.

My first exposure to music was the radio and family members playing vinyl albums in various musical styles. I heard an array of popular music ranging from early rock, British Invasion bands, and folk singers. At age 10, I asked for piano lessons, which engendered my awareness of the varied dynamics and phrasings in musical passages. When my musical interest shifted to guitar at 15 years old, guitar-centric classic rock artists such as Led Zeppelin, Jimi Hendrix, and Stevie Ray Vaughn became appealing. My musical interests expanded further as newer styles of rock emerged, including punk, new wave, alternative, emo, grunge, and indie, among others. At first, the song melodies, lyrics, rhythms, sound, and attitude of the recordings attracted my attention. Later, with more experience as a musician and later as a recording engineer, I began to understand how the artist, engineer, or producer created those sounds (distortion, tape delay, various effects, backwards recording, chorusing, flanging, doubled vocal or instrument tracks, and other studio tricks). The insight from listening to and playing music informs the skills required to evaluate a mix and select a mastering approach. If elements are not placed effectively in the soundstage, a new mix is in order before time and money is spent mastering. This may mean a re-mix for the client, but if the result is a better final master, it is worth the discussion.

Become Knowledgeable About Audio Mastering Tools

The Primary Colors of Mastering

You must become expert at the fundamental tools and techniques of audio mastering, specifically: equalization (EQ), compression and limiting. I refer to these three processes as *The Primary Colors of Mastering* (covered further in Chapter 4—Fundamental Mastering Tools and The Primary Colors of Mastering). Competence with these three represents the basic tools required to begin effectively mastering. Once a command of these fundamental tools is firmly achieved, delving into additional processes will make sense and present options for you to further enhance audio. These *advanced approaches* initially include adding multiple instances of different equalizers or compressors for additional frequency bands or curves (with EQs), tube/transformer coloration from the circuit design, gain staging, or different attack/release times and ratios (with compressors). Beyond that lies *advanced approaches* such as mid-side processing, parallel processing, serial compression or limiting, saturation, clipping, expansion, and multiband compression, among other creative options. These devices and processes are discussed in more detail in Part V: Professional Mastering Process—Micro Considerations.

Electronic Components and Circuit Design Topologies

You don't need to be an electronics technician to be a Mastering Engineer, but it is helpful to have an understanding of the electronic components and circuit design options commonly used in audio equipment due to their sound characteristics. The audio fidelity of equalizers, compressors, limiters, power amplifiers, consoles, and (generally) any audio signal path is affected by the type, quality, and arrangement of their internal electronic components. Hence, the prevalence of descriptive adjectives among audio professionals like: transformer-y, tube-y, solid-state, clean, warm, crisp, transparent, or dark sounding.

Common audio electronic components that are prevalent in compressors, equalizers, consoles and other mastering equipment are: resistors, capacitors, transistors, diodes, inductors, tubes and transformers, and integrated circuits (ICs). These electronic components are arranged into circuits designed with different *topologies* that are associated with varying sound qualities (Figures 1.4, 1.5). These *topologies* fall into

Figure 1.4 The Manley Variable-Mu™ compressor—popular in mastering studios—utilizes tubes, transformers, resistors, and capacitors in its topology. These components impart coloration to program material.

Source: (author collection)

Figure 1.5 Integrated circuits (ICs) have electronics etched onto a silicone substrate and are small and inexpensive. Without including discrete components in the signal path, equipment that uses ICs may yield a generic sound.

Source: (courtesy Capitol Engineering)

fundamental either/or categories: hand-wired or printed circuit boards (PCBs), discrete transistor or op-amps, tubes and transformers, or solid-state. For example, this means that solid-state devices designed with PCBs and IC op-amps can be production-manufactured and are less expensive to build and generally add less coloration or character to the audio signal. Alternately, audio equipment that is: hand-wired, using discrete transistor amplification, and tubes and transformers in the circuit design is labor-intensive and expensive to build and can impart additional desirable coloration to the audio. Coloration is ideally a desirable sonic quality such as: a compressor that extends high frequencies and can be used as a de facto equalizer, an equalizer with a warming or softening effect due to internal tubes, and/or transformers.

Know Audio Frequencies

You must completely memorize the relationship between audio frequencies and musical instruments. This naturally occurs with time invested working on recordings, and is fundamental to your function as a Mastering Engineer. Ideally, your familiarity with sound evolves such that frequency-to-instrument relationships become second nature. This includes the frequency range of instruments. The E.J. Quinby *Carnegie Hall Musical Pitch Relation Chart* from 1941 represents a great reference for this, and elucidates the fact that many musical instruments (and their overtones) encompass a broad band of frequencies worth exploring. However, as a recording, mix, or Mastering Engineer, you will quickly learn that instruments cannot effectively occupy the same frequency range or spatial position in a mix without one instrument being obscured or masked. In order to create an interlocking relationship between these instruments, you must choose a *prominent frequency* to boost or a surrounding *accommodating frequency* to cut in order to place them into the *image* effectively. Obviously, the song key affects the frequencies that will boost or cut an instrument, but Table 1.1 breaks down frequency-to-instrument relationships by *prominent frequency*.

When mixing, the instruments and their *prominent frequencies* and spatial relationships should fit together like an audio puzzle, interlocking to structure the audio image. Panning is another mixing function used to separate and create space among mix elements.

Understand Panning

Panning specifically does not take place in stereo mastering, but you must be aware of how mix elements are panned within the *image*, and why. In mastering, you will be enhancing or reducing the perception of these elements. Although there are ultimately no rules in the art of recording and mixing music, in modern stereo music, certain panning relationships regularly exist (Tables 1.2–1.4). It is beneficial to memorize these, as it will help in your study/analysis and enhancement of the intended image of the recording.

If you have recorded and panned a live drum set as indicated, for instance, it's fantastic to press the mono button on your monitor section, and then switch back to stereo to familiarize yourself with the potentially expansive imaging of the drums. Verify that mix elements with similar *prominent frequencies* are panned opposite or nearly opposite each other in the stereo *image*.

If you receive a mix with the affliction of competing elements, send it back with notes for a revised mix—the master will sound much better if you do so. Stereo audio creates a remarkably pleasing listening experience, but it must be approached with care for maximum effectiveness and correct imaging.

Table 1.1 Common instruments and their prominent frequencies. This information is valuable to isolate and focus EQ moves.

Instrument	Prominent Frequency
Sub-Sonic	18–36Hz
Kick Drum	36–72Hz
Bass	100–400Hz
Guitar	1.4kHz and/or 3.3kHz
Keys	350–650Hz
Vocal Body	240–600Hz
Vocal	700–800Hz and/or 4–6kHz
Vocal Air	15–21kHz
Snare Drum	220Hz and/or 4.8kHz
Cymbals	10–15kHz
Supersonic	21kHz and above

Tables 1.2–1.4 Study these traditional panning relationships; they will help with understanding stereo imaging.

Drums (either drummer or audience perspective):

Instrument	Panning	Note
Kick Drum	Center	Anchors the rhythm
Snare Drum	Slight Left	*Drummer Perspective* FBO of 'air drummers'
Hihat	Hard Left	Cymbals—*Drummer Perspective* indicated here. Often recorded as stereo overheads. *Stereo Array* in alignment with chosen perspective (drummer or audience)
Ride	Hard Right	
Crash(es)	Between Hihat and Ride or Hard Right and Left	
Hi Tom	Hard Left or 9–11 o'clock	*Stereo Array* in alignment with chosen perspective
Mid-Tom	Between Hi and Lo Tom	
Lo Tom	Hard Right or 1–3 o'clock	
Room	Hard Left/Right if stereo	*Stereo Array* in alignment with chosen perspective

Rhythm Instruments:

Instrument	Panning	Note
Bass	Center	Anchors track along with kick
Rhythm Guitars	Hard Left/Right if stereo	In array with other rhythm instruments if mono
Keyboards	Hard Left/Right if stereo	In array with other rhythm instruments if mono

Vocals:

Designation	Panning	Note
Lead Vocal	Center	Anchors track along with kick
Harmony Vocals	Hard Left/Right if stereo	Slightly off-center if mono
Backing Vocals	Hard Left/Right if stereo	In array with other vocals if mono

Know Loudness Metering/Level Measurements

The *loudness*[10] of music is ultimately subjective, and the objective measures of *loudness* do not always correspond to its subjective or *perceptual measures* (aka *apparent volume*). The Mastering Engineer's ears are the ultimate arbiter of appropriate *loudness*. To support this critical aspect of audio mastering, there are several types of *loudness metering* options, but you will need at least two: an analog volume unit (VU) meter (with adjustable 0VU *reference level*), and a decibels full scale (dBFS) meter (often onboard your *digital audio workstation* [DAW]). Plug-in versions of these meters are also readily available.

A VU meter is a *relative measure* corresponding to a voltage reading in *decibels unterminated* (dBu) so that 0VU = .775V = 0dBu. Most professional analog recording consoles are fitted with VU meters to allow the engineer to set the *gain structure* of a mix. The standard VU meter alignment for an analog console is 0VU = 1.23V = +4dBu. The same is true for a traditional mastering console (or equipment chain)—a set of VU meters were included, and as master levels increased with the advent of digital recording, the 0VU *reference level* on the VU meters could also be increased to display peaks without pinning the meters. For most styles of music, an effective VU meter reference setting for mastering is 0VU = +12dBu.

Take note that this setting shows audio peaks that are a full 8dB higher on the VU meter than the standard mix level of +4dBu described earlier, which indicates the amount of gain routinely added in mastering. The number of mixing or Mastering Engineers I encounter who do not use VU meters in their work remains a mystery to me. I regularly advise mixers that if they get accustomed to using VU meters, the quality of their mixes will exceed their competition in gain structure, instrument balances, and impact—and the same applies to mastering. There are only benefits to using VU meters (Figure 1.6) in a mastering system, and PSP even makes an effective and inexpensive plug-in, the VU3.

A dBFS meter makes *absolute measures* with each unit representing 1dB. This measurement, along with the VU meter reading, is important to assess both the flat mix and also the mastered result. At the RDAW, where mastered audio is captured or *zone 3* of your mastering system, you will need a reliable dBFS meter. The first thing to remember is that you cannot exceed 0dBFS without clipping or distorting the audio, and the dBFS meter will indicate as much by showing red over-levels. Also, you must align the digital domain(s) of your mastering system—via the voltage level at DA conversion—to a standard analog *reference level* at the VU meter. There are several options for this, but I adhere to $-14\text{dBFS} = 1.23\text{V} = 0\text{VU} = +4\text{dBu}$ (using a 1kHz tone from the PBDAW), and it has been reliable and effective for all genres of music. Other common *reference level* alignments for the PBDAW DA converter output are -16dBFS or -18dBFS (with makeup gain to 1.23V at the DA converter) $= 0\text{VU} = +4\text{dBu}$, which limits headroom below 0dBFS, allowing more dynamic musical genres such as classical, solo instrument, or jazz to achieve a robust level.

Again, good VU and dBFS Meters are a minimum meter-wise to master effectively. The EBU r128[11] standard for broadcast loudness references three additional *metering scales* that are valuable to implement, *loudness units relative to full scale* (LUFS), *loudness K-weighted full scale* (LKFS),[12] and *loudness units* (LU). An LUFS/LKFS reading is an *absolute digital measurement* where each unit is equal to 1dB. It is similar to an average or *root mean square* (RMS) measurement relative to dBFS—also an *absolute digital measurement*. Spotify and YouTube, among other digital streaming/distribution platforms, list a target LUFS level for audio submitted to them (–14LUFS and –13LUFS, respectively), so it is valuable to notate this measurement, as well. LU is a *relative analog measurement* similar to an analog VU meter voltage reading, displayed as a positive number. LUFS and LU levels can also be notated from song to song to verify cohesion for an album or other collection of mastered songs. In classic Mastering Engineer meticulousness, I regularly notate up to five *loudness measurements* for each song while mastering an album: VU, dBFS Peak, dBFS RMS, LU, and LUFS.

Figure 1.6 Hardware VU meter which is relative to voltage of .775 V at 0VU. This custom-made version by Capitol Engineering for my mastering studio has an adjustable 0 reference switch to properly read VU peaks of loud masters.

Source: (courtesy Capitol Mastering)

Become Knowledgeable About Acoustics and Listening Environments

The Effect of Acoustics on Sound

Spend time listening to live musical performances that encompass a variety of genres. Attention to acoustics at concerts in quality performance spaces will create a parallel awareness for you to reference in the mastering studio. This will also inform you of the desirable effects of room dimension and surface material on sound. As described previously under *deliberate listening*, awareness of what is pleasing or displeasing will inform your mastering work. Protect your ears in performance spaces, so always have ear plugs on hand in case the sound reinforcement is excessive.

Ideally, both performance venues and studios are spaces designed and engineered to effectively present a full-frequency range, uncolored *listening experience*. Here in Los Angeles, Frank Gehry's Walt Disney Concert Hall, with the interior hall designed by the triumvirate of Gehry, acoustician Yasuhisa Toyota, and conductor Esa-Pekka Salonen, represents a notable example of a meticulously designed, wood-appointed concert hall. Having attended a few LA Philharmonic performances there, my impressions were that the hall is diffuse enough for a consistent listening experience regardless of seating position, and the unamplified acoustics are detailed in presentation, balanced frequency-wise, focused image-wise, and rich in general character, and transients/dynamics are not artificially excited, with the high-frequency extension slightly curtailed. Initially, the orchestra required extra rehearsals in order to adjust to the revealing acoustics, and conductor Salonen even noticed longstanding notational errors in the score for Ravel's *Daphnis and Chloé*.[13] Different examples include renowned Los Angeles outdoor venues such as the Hollywood Bowl and The Greek Theater which have the largely non-reflected sound characteristic of open-air venues, supported by substantial sound reinforcement at the front and sides of the venues, helping audio hold together well despite sound escaping into the air with no walls or ceiling to reflect or contain it. I cite these examples to reinforce that the size, shape, and dimensions of the venue, along with the *acoustic properties* of its various surfaces, including any acoustic treatments, affects *listening experience*.

Acoustics and listening environments affect the presentation of music and sound. Acoustically treated environments are necessary for accurate listening assessments. A grasp of the natural sound of musical performances, as well as the effective design and treatment of the mastering studio allow the Mastering Engineer to make appropriate decisions and adjustments for an excellent sounding master recording. In Chapter 3—The Mastering Studio, I will further explore relevant aspects of a controlled listening environment

Become Proficient With Digital Audio Workstations (DAWs)

The optimized and mastered audio is captured (recorded) and rendered into the required delivery formats and resolutions in a DAW. This is simply a computer with an audio input/output card and software that can play back, edit, process, and record audio—and also generate all delivery formats (standard or high-resolution .wav files, Red Book PMCD Masters, or DDP CD Masters). I recommend and implement a two-DAW mastering system, one dedicated to playback (PBDAW) and the other dedicated to capture/record (RDAW). In Chapter 4—Fundamental Mastering Tools and The Primary Colors of Mastering, I will review the pertinent aspects of DAWs in the mastering system.

Become Adept at Quality Control (QC) Methods

My foray into audio mastering began as a *production engineer* at Capitol Records, where my main function was QC. This represents a deceptively simple, but extremely crucial and critical skill that every Mastering Engineer must possess, also known as . . . catching mistakes. *QCing* involves verifying the integrity of all aspects of a master, and falls into two main categories: integrity of the audio fidelity, and release information and metadata. Among other things, this means listening to the entire master (album, EP, or single) to make sure it is suitable for distribution or replication. All compact disc (CD) masters include release-specific information and metadata that must be absolutely 100% correct to the numeral and character. I will explore QC further in Chapter 9— Step V: Delivery—Generate the Master.

Become Able to Generate the Audio Master

Once the mastered audio is approved, you must generate and deliver the requested formats for the client. My approach is to capture a high-resolution mastered .wav file in my RDAW (usually at 96kHz—32bit floating-point) as a raw *archive master*. I then edit tops/tails or assemble the project, dither, and render to all the subsequent required delivery formats. These are usually either a CD Master (PMCD or DDP), .wav or .aiff files, or vinyl cutting files. Obviously, with *high-resolution mastering* above 96kHz, if 192kHz is requested, I make the initial capture at 192kHz.

Implement Recallable Workflow Approaches

Mastering Notes and the Capitol EQ Rundown Card

At Capitol, going back to the late 1940s or early 1950s when the transition from transcription discs to analog tape was occurring, engineers would fill out and attach an EQ/ Rundown card (Figure 1.7) to the tape box for documenting EQ, level, compressor, or other equipment settings that were used to create the master. This card would get taped onto the flat compiled master tape—this way a new production master could be made if need be, as even safety or backup copies could get damaged or, even worse, lost. I developed a habit of using these cards to notate processing settings for every mastering project I do, and I now have thousands of them organized by year in a number of boxes. So, you must have mastering notes for each song or project you master. This represents a critical habit for the following reasons:

1. Your client may add songs or request alternate versions of the same song.
2. Your client may ask for a revision and you can recall and adjust as needed.
3. Your client may call a year or two (or more) later, having loved the earlier result. You can exactly re-create your signal path and match levels and fidelity.

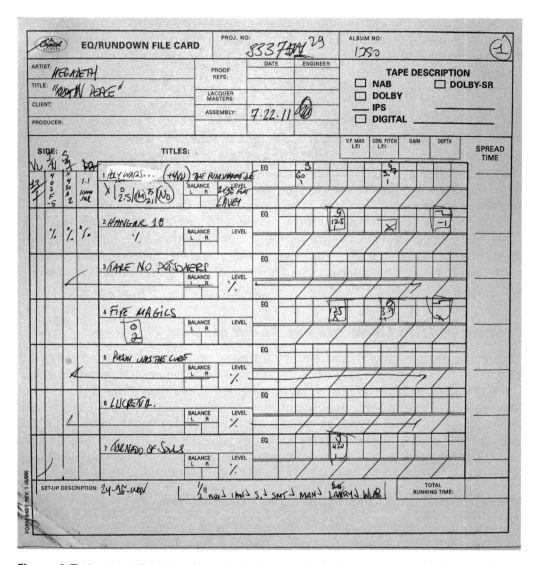

Figure 1.7 The Capitol EQ Rundown Card. My standard documentation methodology, I have thousands—one for every mastering project.

4. You can keep track of the projects you work on, in case they 'blow up' (go gold, go platinum, chart, get press, get downloads, etc.).
5. You can keep track of your network of artists, bands, producers, engineers, and record labels for marketing strategies or analysis of your mastering work.

Notes on Recalling Settings

Analog equalizers, compressors, and limiters designed for mastering usually have *stepped attenuators* with *precision resistors* at each position so that the adjustments are perfectly balanced between left and right channels (preserving correct imaging), and also so that settings can be recalled exactly. Some equipment manufacturers use a potentiometer with a *wiper* that moves across a single carbon track, which is less reliable than a different *precision resistor* value at each gain/EQ/bandwidth setting. One of the great advantages of the plug-in era is that your settings can be quickly saved in the plug-in folder for immediate recall.

Naming Conventions

You must adopt a naming convention to organize your projects, files, revisions, and notes so that they provide relevant information at a glance, and you can find them quickly if required. My naming convention is as follows for audio projects and files. I *always begin with the year*.

> *PBDAW Audio Drive*
> Main Folder: Year
> > Nested Folder 1: Year_Artist_Album Name_S-BD
> > ProTools Session and Folders with Source Files Here
> *RDAW Audio Drive:*
> Main Folder: Year
> > Nested Folder 1: Year_Artist
> > > Nested Folder 2: Album Name
> > > > Raw Mastered Files Here: Song Title_S-BD
> > > > Nested Folder 3: Mastered WAV Files_44.1–16
> > > > Nested Folder 3: Mastered WAV Files_MFiT
> > > > Nested Folder 3: Mastered WAV Files_Vinyl

The PBDAW has a more detailed *Nested Folder 1*, as ProTools creates an encapsulating *session folder*, whereas WaveLab on the RDAW does not, and usually has the same S-BD (96kH–32bit floating point) for most projects.

Become Personable and Entrepreneurial in Demeanor

Achieving proficiency in audio mastering, along with excellent results, remains central to a professional Mastering Engineer. In addition, authentically connecting with artists, engineers, producers, and record label personnel should not be overlooked. Being personable and representing a quality mastering solution can help keep you in consideration among many other mastering options. Mastering represents a very competitive arena, and technology has at best democratized, and at worst eroded, understanding and respect for the discipline. The flow and ingestion of projects is determined by word-of-mouth discussions between artists, producers, or label representatives where your name and good reputation ideally circulate. Ultimately, if you work for yourself, an independent mastering studio, or a major label corporation, you must create an interest in and a demand for your mastering work by connecting with music creators, delivering excellent results, and working on projects with cultural and/or artistic resonance. It is a big world, and many markets exist that seek a Mastering Engineer who understands the nuances of their music, and can help deliver superlative audio fidelity within their budget.

KEY UNDERSTANDINGS TO POSSESS

Sampling Theorem

Attributed to Harry Nyquist and Claude Shannon, *sampling theorem* states:

> continuous-time (analog) band-limited signals can be replaced by discrete time (digital) samples without any information loss. The original continuous signal can be reconstructed from the samples. The sampling frequency (S) must be at least twice the highest signal frequency. The digital system determines the numerical values it will use to represent waveform amplitude at each sample time (quantization).[14]

Understanding *sampling theorem* and how digital audio principles affect audio fidelity remains essential knowledge to a Mastering Engineer and informs preliminary analysis of a digital source file, which I refer to as *objective assessment* (explored in greater detail in Chapter 5—Step I: Objective Assessment). The *sampling frequency* (S) is how many times per second the analog-to-digital (AD) converter samples the analog signal. *Sampling frequency* correlates to audio bandwidth by means of the *Nyquist frequency* (S/2, or half the sampling frequency) (Table 1.5). This means that a file with a *sampling frequency* of 44.1kHz will have a bandwidth of 0Hz–22.05kHz.

Bit depth (BD) pertains to dynamic range at a relationship of 6dB per bit (Table 1.6).

These two parameters determine two critical limits of the audio fidelity of a digital file: its bandwidth and dynamic range. If you are boosting a very high frequency of 25kHz into a 44.1kHz recording, that recording will not reproduce any high frequencies above 22.05kHz, but a 96kHz recording will (provided that your playback components and speakers can reproduce those frequencies). In the zeal for high-resolution audio, we conveniently forget that microphones, speakers, *analog-to-digital*

Table 1.5 The relationship between sampling frequency (S) in kilohertz and audio bandwidth. The upper frequency is known as the Nyquist frequency (N), or half the sampling frequency (N=S/2).

Sampling Freq. (kHz)	Bandwidth (Hz–kHz)
44.1	0–22.05
48	0–24
88.2	0–44.1
96	0–48
192	0–96

Table 1.6 The dynamic range of 16bit, 24bit, and 32bit floating-point digital words at 6dB per bit. Note that a 32bit floating-point digital word includes a 24bit mantissa and an 8bit exponent.

Bit Depth	Dynamic Range (dB)
16	96
24	144
32fp (24bit audio)	144

and *digital-to-analog* (ADDA) converters, certain plug-ins, or other audio processing equipment might not reproduce ultra-high or ultra-low frequencies that are incidentally beyond the human hearing range of approximately 20Hz–20kHz. It is easy enough to test and hear that 24bit resolution digital audio produces a richer more dimensional sounding recording than 16bit digital recordings; 32bit floating-point is a great high bit rate storage format for digital audio. Many DAWs operate using 32bit floating-point *internal processing precision*; this is a good match between archived audio (that may need digital signal processing or DSP downstream) and the computer's processing precision. Also remember that the Mastering Engineer must render multiple formats of digital audio files. For these reasons, I advise that you master at 96kHz—32bit floating-point. This allows you to make DSP adjustments later with the highest quality, and also allows you to make 96kHz—24bit high-resolution audio files and also *Mastered for iTunes* (MFiT) compliant files, along with 44.1kHz—16bit *standard digital* audio files.

Practice listening for the fidelity differences between *lossy* digital files (.mp3, .aac, .m4a), and *lossless* digital files (.wav, .aiff or .flac). Further, you should practice listening to the difference between different resolutions of lossless files. This means different sampling frequencies, and bit depths, mostly 16bit versus 24bit, and 44.1kHz, 48kHz, 88.2kHz, 96kHz, 176kHz, and 192kHz. Note that due to the extreme bandwidth of high-resolution audio, and the limitations of human hearing, differences can be impossible to hear. This is due to the fact that the *Nyquist frequency* of high-resolution audio can extend to 96kHz or beyond (in the case of a 192kHz or higher sampling frequency). Although technology has superseded the bandwidth of human hearing, proponents cite better signal-to-noise ratios and visceral responses to high-resolution audio. Certainly, *digital signal processing* (DSP) benefits from the interpolations taking place at higher sampling frequencies than at standard 44.1kHz.

Analog Domain vs. Digital Domain Considerations

Mastering adjustments in just the analog domain or just the digital domain yield different sounding results that both have merits. Analog equalizers may not have the surgical parameter adjustments found in plug-ins or standalone DSP hardware (i.e. Weiss EQ1 and DS1, Waves L2, or t.c. electronic M6000). Conversely, the digital options may not impart the dimensionality, tonal character, coloration, or even saturation and clipping of great analog mastering equipment. After years of experimenting with different processing chains, my favorite approach is a *hybrid chain* with both digital processing and analog processing occurring in the *mastering system*. There are certainly situations when I favor a primarily all-analog chain or an all-digital chain, but I consistently return to the hybrid setup for the best overall audio fidelity.

CONCLUSION

This chapter explored the multiple facets, understandings, and expertise required of a modern professional Mastering Engineer. It is valuable to identify the many component areas of knowledge that make up the greater discipline of audio mastering. It provides a bird's-eye view of the discipline for an artist, producer, engineer, or enthusiast considering mastering as a career. It also provides a practical checklist for any practicing Mastering Engineer to assess strengths or weaknesses. Additionally, I began to lay out specific approaches and concepts that make up my Major Label Mastering Process.

EXERCISES

1. Select two competencies of a Mastering Engineer from the list of ten described in this chapter. Explain why they are important. What problems can a Mastering Engineer manage or solve with the two competencies you selected?

2. Sampling Theorem describes the relationship between which two domains? How do sampling frequency and bit depth pertain to the analog audio signal? What is the Nyquist frequency of a 48kHz digital file?

NOTES

1. Shunryu Suzuki (2010) *Zen Mind, Beginner's Mind: Informal Talks on Zen Meditation and Practice*, Boston, MA: Shambhala.
2. Robert Harley (2015) *The Complete Guide to High-End Audio, 5th Edition*, Carlsbad, CA: Acapella.
3. Francis Rumsey (2004) *Desktop Audio Technology*, Burlington, MA: Focal Press. pp. 52–54.
4. Bobby Owsinski (2005) *The Recording Engineer's Handbook*, Boston, MA: ArtistPro Publishing. pp. 62–63.

    Blumlein Array—Developed for EMI in 1933 by famed audio pioneer Alan Blumlein, the Blumlein stereo setup is a coincident stereo technique that uses two bidirectional microphones in the same point and angled at 90 degrees toward each other. This stereo technique will normally give the best results when used at closer distances to the sound source, since at larger distances these microphones will lose the low frequencies. The Blumlein stereo has a higher channel separation than the X/Y stereo but has the disadvantage that sound sources located behind the stereo pair also will be picked up and even be reproduced with inverted phase.

5. Alan P. Kefauver, and David Patschke (2007) *Fundamentals of Digital Audio, New Edition*, Middleton, Wisconsin: A-R Editions. p. 25.
6. Angela Duckworth (2016) *Grit: The Power of Passion and Perseverance*, New York, NY: Scribner. pp. 120–142.
7. Robert Harley (2015) pp. 2–3.
8. Ken C. Pohlmann (2011) *Principles of Digital Audio, 6th Edition*, New York, NY: McGraw-Hill. p. 336. "Psychoacoustics must also consider the perceptual measurements of an audio event . . . loudness cannot be empirically measured and instead is determined by listeners' judgments."
9. F. Alton Everest, and Ken C. Pohlmann (2015) *Master Handbook of Acoustics, 6th Edition*, New York, NY: McGraw-Hill. p. 46. Seminal work on loudness was done by Fletcher and Munson at Bell laboratories in 1933.

10. tc electronic (2019) *Loudness Explained*, Riskov, Denmark: tc electronic. See www. tcelectronic.com/brand/tcelectronic/loudness-explained#googtrans(en|en).

11. See EBU r128 white paper "Loudness Normalisation and Permitted Maximum Level of Audio Signals," at https://tech.ebu.ch/docs/r/r128.pdf

12. Loudness K-weighted Full Scale uses a K-weighted filter curve which bridges subjective impression with objective measurement to help measure perception of loudness. LKFS is effectively synonymous with LUFS.

13. Mark Swed (October 29, 2003) "Now comes the true test." *Los Angeles Times*.

14. Ken C. Pohlmann (2011) pp. 20–25, 731–734.

CHAPTER 2

Listening Experience

The Mastering Engineer's Primary Concern

Audio mastering embodies a discipline concerned with enhancing *listening experience*—how the music affects the listener viscerally, allowing them to experience it inclusive of its attitude, message, melodies, and rhythms. The Mastering Engineer's adjustments to the audio and their sensitivity to the underlying song sections and performances ideally elevate the listener's sonic journey. In that regard, final masters should possess, and be checked against, The Eleven Qualities of Superb Audio Fidelity: 1) level/apparent loudness, 2) image, 3) transients, 4) depth, 5) definition, 6) clarity, 7) detail, 8) extension, 9) correct dynamics (*micro-dynamics* and *macro-dynamics*), 10) vocal halo, and 11) blossom. Effective mastering improves these audio characteristics. These qualities represent an invaluable 'checks and balances' list to reference, so practice identifying them in your mastering work. I'll explore these qualities in more detail following.

THE ELEVEN QUALITIES OF SUPERB AUDIO FIDELITY

Level/Apparent Loudness

Level and/or *apparent loudness* remain primary audio mastering concerns. As discussed in Chapter 1 under competency #4—Become Knowledgeable about Audio Mastering Tools, *level* is objective, and *loudness* is subjective or perception-based. *Loudness* is also inversely proportional to the overall *dynamic range* of the recording, and making a recording excessively *loud* often presents a quality trade-off. How *loud* or 'hot' should a professional master be? There are many considerations that inform the answer, including genre-based and market expectations, impact of the production, and artist preference. In the commercial music business 'make it hot' remains a common refrain by artists and repertoire (A&R) executives, as record labels believe that it contributes to the hit potential of a recording. If you can create clean (artifact-free), impactful masters that compel listening, chances increase that you will have a line of customers outside your mastering studio.

A fundamental indicator of mastered audio *level* and *loudness* is your playback volume setting. If you audition a flat mix or dynamic recording, it requires a higher volume setting to achieve the *sound pressure level* (SPL) you are accustomed to hearing.

Conversely, if auditioning a modern peak-limited recording, the volume will remain lower to achieve a similar SPL since the track has less *dynamic range* and more *apparent volume* (aka *loudness*). In addition, awareness of the *target levels* of the various genres of music you are working on will frame your approach to the level of audio masters. Thus, it remains beneficial to keep current popular recordings in your DAW to reference and measure so that your masters are competitive level-wise with market trends and expectations. Conversely, don't unwittingly enlist in the *loudness wars*: you may conclude that a lower level suits the recording best.

Image

The perceived space between the left and right speaker in stereo audio configurations is the *image*. The placement and presentation of instruments within that space is referred to as *imaging*. I often visualize the Hollywood Bowl as a great pictorial representation of *image* because as opposed to a flat plane, it includes *width*, *height*, and *depth*. Mix elements panned hard left and hard right create *width* and an expansive image—generally desirable. Mix information emerging from the center of a stereo playback system establishes the *phantom center*, as there is no speaker there, but audio is heard from the center, directly in front of the listener. If a mix is recorded in mono, the *imaging* is limited and instrument and frequency balances take precedence.

Height conceptualization in the recorded audio *image* is informed by a number of observations. Higher frequency instruments emerge from tweeters or super tweeters, which normally sit above a woofer or mid-range driver in a speaker cabinet. Considering a drum set, the low frequencies of a kick drum (physically on the floor) appear to anchor the *image*, and cymbals remain higher, on stands. A vocalist presents upper mid-range frequencies at head height. These are some of the reasons why high-frequency information is perceived at the top, and lower frequencies at the bottom region of the *image*. *Depth* (item 4 ahead) represents another important component of the *image*.

An excellent master will make use of the space within the *image*, and even correlate *imaging* with the dynamics and song sections of a recording. For example, an introduction or verse may have a more narrow or shallow *image*, but with the entrance of other instruments or new song sections, the *width* and *depth* of the *image* should become more apparent as well.

Transient Response

This is the initial impulse or attack from an instrument or sound source. *Transients* are readily apparent on drums, percussion, and plucked instruments, but in general every instrument and sound has transient response, and it lends vitality and excitement to music and recordings. It is advantageous to preserve or even enhance *transients*. Remember that compressors and limiters directly lower *transients* to avoid distorted peaks or over-levels, or are also used to increase the average volume of an instrument or mix. Conversely, an expander can enhance *transient response* in a mix. *Transients* will be the first place that a master will begin to distort, informing you that you are pushing the level too much. One aspect of a great master is to hear and feel the vitality of *transient* attacks throughout the music. This often begins with the mix engineer providing a dynamic and well-balanced mix.

Depth

This describes the three-dimensional perception of front-to-back instrument placement within the *image* and refers to the perceived distance of the sound source from the microphone—and hence, the listener. Effective mastering will enhance the presentation of depth, which is largely adjusted at the mix stage. In the Hollywood Bowl analogy, there is a front and back to the soundstage, in addition to numerous left and right spatial positions that create distance from the listener. With the dimension of *depth*, you can alternately envision the *image* as a snow globe. When you hear an orchestra, jazz trio, or rock band perform, members are positioned on the stage as a function of their role in the ensemble or to balance and blend the characteristics of their specific instrument. In the orchestra example, the percussion section is placed at the back of the stage, then brass, with woodwinds and strings positioned progressively toward the front. Soloists are routinely featured in front across most genres of music. Enhancing depth in your masters will contribute to an enhanced *listening experience*.

Definition

This indicates the discernable outline of each element within the *image* as a whole. The opposite would be an undefined smear between recorded elements—a masking of sounds with similar frequency ranges. Skilled parallel compression is an excellent way to add definition in mastering.

Clarity

This refers to the unobscured and lifelike immediacy of the master. Ideally, it presents as if the musicians are in the room with you. Each musical element has a defined spatial position in the image, and some discernable space between them. A master with clarity sounds natural and has nothing in the way of artifacts or noises (hiss, distortion, or other anomalies). The opposite would be a dull or slightly obscured mix presentation.

Detail

This refers to each instrument's discernability, including nuances of timbre and complexities of overtones or harmonics. For instance, a snare drum has an attack transient and a broadband tonality which includes the characteristic of the drum head, rattle of the snares, the material the drum shell is made of (usually metal or wood), and sustain, along with any correlated reverberation. Or, a vocal possesses breath, chest, and/or head resonance, tone, diction, vibrato, and any correlated reverb or delay. Practicing listening to solo instruments will inform your awareness of *detail*. *Detail* is a desirable component of a great master, and Mastering Engineers regularly seek to enhance this aspect of a mix.

Extension

This pertains to frequency response and refers to the openness of the high frequencies, as well as to the feel and resonance of low or sub-harmonic frequencies. It may or may not be appropriate for all genres of music, but it is generally very appealing. *Extension* can be identified in the entire mix, but also in elements such as vocals, backing vocals, broad-range instruments, cymbals, kick drums, bass guitar, and synthesized bass sounds. It is the opposite of a 'band-limited', contained or mid-range sound quality. To add *extension*, use hi-shelving or low-shelving EQs with increasing slopes (i.e. tilt, classic Pultec or Gerzon shelf), or a bell-shaped EQ curve set above 20kHz or below 10–20Hz (see Chapter 4—Fundamental Mastering Tools for EQ details).

Correct Dynamics (Musicality)

Dynamics refer to the range of amplitude in a recording. The term applies to both the moment-to-moment differences between transient peaks and subsequent sustain or decay of instruments and music (*micro-dynamics*), and also the changes in intensity between song sections within the recording (*macro-dynamics*). The sonic journey of a recording marked by melodic and rhythmic changes is ideally supported by these dynamic changes, and a great master preserves them despite level adjustments and compression. *Dynamic range* can be measured in the digital domain by subtracting the *decibels full scale* (dBFS) Peak level from the *root mean square* (RMS) or average dBFS measurement of the program material. This yields the *dynamic range* in decibels. Alternately, taking VU meter level readings in the analog domain in dBu and subtracting the lowest reading from the highest reading would also indicate the *dynamic range* of the recording. Referencing both measurements is ideal when evaluating both flat mixes and final masters.

As mentioned in the previous paragraph, *dynamics* are classified into two subcategories, *macro-dynamics* and *micro-dynamics*. *Macro-dynamics* are the changes in level between sections of a musical piece. For instance, introduction to verse to pre-chorus to chorus; or, in the case of a symphony: allegro, adagio, scherzo, sonata. These sections relate to one another, and there is an expectation that the chorus section or allegro movement remain louder and busier than the verse section or adagio movement. In most instances of pop, rock, or rap music, the 'payoff' is in the climax of the chorus, usually bigger and louder than other song sections. Over-limiting can cause the affliction of *inverse macro-dynamics*, whereby the quiet sections are louder than the big sections—definitely avoid this. To clarify further, *micro-dynamics* are the dynamic changes within the music itself, at any given point in the recording, independent of song section.

Vocal (or Featured Instrument) Halo

I refer to *halo* as the way the vocal or featured instrument *presents* in the image. A well-mastered recording showcases the vocal in a *halo* of space, detail, and clarity at the center of the image, often forward and slightly above any accompaniment. Hallmarks of this are clear diction and a breadth of frequencies expressed by the vocal. Mid-Side (M/S) EQ and/or compression can effectively enhance vocal frequencies or even lift the vocal track forward (see Chapter 15—Mid-Side).

Blossom

This refers to the payoff of a chorus, hook, or other hallmark section of a recording. Every section of a song plays a musical role, and most genres of popular music include the climax of a chorus or hook. *Blossom* is a specific aspect of *macro-dynamics* and often occurs when the most mix elements are present. The Mastering Engineer must support this natural build and payoff. In these sections, a slight level increase or the widening of the *image* are methods to enhance *blossom*.

CONCLUSION

If you can understand, relate to, and ultimately enhance The Eleven Qualities, your masters will reflect a high degree of audio fidelity and positively impact *listening experience*. These characteristics provide a checklist with which you can evaluate the mix and your adjustments while mastering.

EXERCISES

1. Practice listening to and noticing stereo *image* by listening to a stereo recording and switching from mono to stereo on your monitor controller.
    a. Notice how the breadth of the *image* changes while in stereo, and the degree of phase cancellation of stereo information while in mono. Which mix elements do and do not move or shift in either configuration? Which elements constitute *phantom center*? Notate observations.
    b. Identify four mix elements by orientation in the *image* using an imaginary clock dial superimposed on the image.
    c. Determine whether the drums are *drummer perspective* or *listener perspective*, and explain the panning clues.

2. Practice listening for and identifying leading edge *transients* in two versions of a recording (one dynamic and the other peak-limited).
    a. Notate your observations of the relationship between dynamic range and *listening experience* in a recording.
    b. Explain how preserved *transient response* or 'flattened transients' from peak-limiting affects the *apparent volume* of the recording.

3. Practice listening to over-compressed/limited audio and notice how it affects *transients*, *macro-dynamics*, and low-mid and high-frequency information. Notate your observations.

4. Select and compare two examples from the lists in parentheses below. Listen and make observations considering The Eleven Qualities discussed in this chapter. Notice how compression, *transient response*, and *level* are represented in the following three categories:
    a. Genres of Music (jazz, classical, rock, singer/songwriter, pop, hip-hop/rap, etc.).
    b. Eras of Music (1940s–50s, 1960s–70s, 1980s–90s, 2000s–present).
    c. Recording Methods and Approaches (live, multi-tracked, layered sounds, sampled sounds, analog recording, digital recording).

The Mastering Studio and Fundamental Mastering Tools

In Part II, I explore the proverbial laboratory of sound wherein the Mastering Engineer is able to accurately assess audio and then decide on the tools and methods for the best-sounding outcome. Mastering Engineers often spend years experimenting with, refining, or customizing their arsenal of equipment and mastering approaches. As such, a clear understanding of common audio treatments, studio components, and fundamental tools is vital for a Mastering Engineer.

CHAPTER 3

The Mastering Studio

A CRITICAL LISTENING ENVIRONMENT

A professional *mastering studio* (Figure 3.1) represents a flat and uncolored *critical listening environment* designed to reproduce full-range audio. It must be purpose-designed and functionally dedicated to the playback, assessment, and processing of audio. The absence of frequency or imaging anomalies allows the Mastering Engineer to confidently make accurate audio adjustments. This also ensures that masters created in the mastering studio *translate* accurately (sound the same or similar on other systems as they did in the studio) to downstream playback systems ranging from ear buds or headphones to car stereos and high-end audiophile systems. This chapter examines essential components of *the mastering studio*.

The Room

Common Design and Treatment Approaches

The ground-up design and implementation of a professional mastering studio, and/or the retrofitting of existing spaces, remains beyond the scope of this book. However, there are functional design concepts of the studio that a good Mastering Engineer must understand. The room that houses the mastering studio is ideally *sound isolated* so that environmental noise does not enter and distract *critical listening*, nor does playback audio escape and affect accurate frequency response or imaging. Room dimension, volume, and geometry represent vital considerations. This results from the physical length (and energy properties) of sound waves, especially at low frequencies—a 50Hz wave is 22.51 feet long, and a 20Hz wave is 56.26 feet long. This means that ideally, your listening room would be *at least* 23 feet long with a volume of at least 2,500 cubic feet to accommodate a 50Hz sound wave.[1] For this reason, and to alleviate other common acoustic anomalies, *acoustic treatments* in the form of *absorbers*, *bass traps*, *diffusers*, and *resonators* are added for the accurate presentation of sound. Following is a brief definition of each.[2]

1. *Absorbers*—Primarily absorb high frequencies. Strategically placed to absorb reflections from studio surfaces—walls, ceiling and floor—to the listening position. Quotidian examples are acoustic tile, carpeting, curtains, mineral wool, and fiberglass. *Clouds* are *broadband absorbers* that hang from the studio ceiling to dissipate and absorb diffracted sound waves.
2. *Bass Traps*—A membrane absorber designed to minimize buildup of low frequencies. Usually placed at the back and in corners of the studio. Low-frequency issues are common due to the long length of sound waves below 50Hz.

Figure 3.1 Mastering Studio 2 at Capitol Mastering. PBDAW (*zone 1*) is left, RDAW (*zone 3*) is center. The Sterling Modular desk holds all analog equipment (*zone 2*), which is in between the AD and DA converters.

Source: (courtesy Capitol Mastering)

3. *Diffusers*—Characterized by an irregular surface geometry. The relationship between these irregularities and the sound waves striking them determine the frequencies that are diffused. Used to create a large diffuse sound field for consistent imaging at various listening positions. Also constructed as an *Absorptive Diffuser* to combine attributes of both treatments.
4. *Helmholtz Resonators*—Size-built air spaces with holes—analogous to a gallon jug resonating—to absorb specific resonant frequencies and standing waves via phase cancellation.

Professional acoustical treatment companies will measure, analyze, and photograph your room in order to generate a plan for recommended treatments to create a flat and focused listening environment (Figure 3.2). GIK, RPG, Primacoustic, and Vicoustic are companies that offer effective acoustical treatment options.

Acoustic Properties and Sound

Room dimensions—along with varying ceiling, wall, and floor surfaces—affect how sound presents in the mastering studio. These surfaces will have one or a

Figure 3.2 Acoustical treatments in Mastering Studio 2 at Capitol Mastering. Dual pane windows behind the main speakers are angled up to reflect sound into absorptive material and frequency-tuned cloud absorbers in the ceiling. Fabric-covered fiberglass on walls manages reflections, and absorbers at incident angles to the speaker drivers alleviate image smearing and comb filtering at the listening position.

Source: (courtesy Capitol Mastering)

combination of three basic *acoustic properties*: *absorption* (the absorption of sound energy which dissipates as heat energy), *diffusion* (the scattering or distribution of sound into other directions), and *reflection* (the sound energy bouncing off of the various studio surfaces). Larger rooms with high ceilings and reflective surfaces create a longer reverberation decay time, and smaller rooms will have a shorter reverberation time. A good acoustician can advise how 'live' the room should be, as it remains a matter of room accuracy and preference. Concrete, brick and glass are *reflective*, and can cause reflections and add reverberation and vitality to music. Conversely, carpets, curtains or fabric are *absorptive* and tend to smooth out mid or high frequencies. Wood generally offers a smooth and warm reflective quality, but wood types and finishes certainly affect presentation of sound. Correlating the range of surfaces and dimensions with the quality of sound presentation will inform both design approaches for the mastering studio and your assessment skills of recorded audio.

Common Acoustical Issues—Standing Waves, Flutter Echo, and Comb Filtering

The mastering studio should be symmetrical left to right so as to present stereo speaker cohesion and accurate imaging. The geometry of the room should be designed to provide reflective and absorptive surfaces at the proper angles to create diffuse and predictable sound fields without problematic anomalies such as *standing waves, flutter echo,* or *comb filtering.*[3] Reflective parallel walls must be avoided at the design stage or treated, as they create *standing waves* whereby certain frequencies are multiplied and resonate as a function of the sound wavelength, and the distances from the sound source to the walls and the listening position. *Flutter echo* is another issue caused by parallel walls, as sound will reflect between them, causing an undesirable repetitive echo sound. *Comb filtering* is caused by the delayed frequencies reaching the ear at slightly different times so that some frequencies are phase-cancelled and others are amplified, resulting in a representation that looks like a comb rather than a smooth curve (Figure 3.3).[4] *Image smearing* describes an acoustic phenomenon whereby aspects of the audio image sound undefined and smeared due to the issues previously described occurring at the listening position. The Mastering Engineer must make accurate decisions and adjustments, so all acoustic anomalies must be rectified in order to achieve accurate presentation of the music. Managing these issues involves adding *acoustic treatments* to the mastering studio that allow for the true and uncolored presentation of full-range audio.

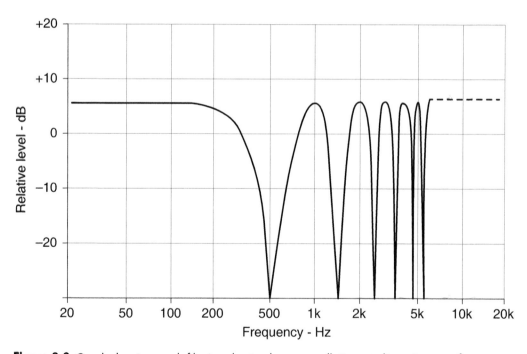

Figure 3.3 Graph showing comb filtering due to phase cancellations at dimension-specific frequencies in a room. An accurate assessment of audio quality cannot be made in a room with comb filtering issues.

Acoustical Testing and Treatments

As discussed previously, some aspects of a room are undesirable. In order to accurately rectify these, acoustic measurements of frequency response are taken by playing *pink noise*[5] through the speakers and placing a microphone connected to a spectrum analyzer at the listening position. This, in conjunction with user listening tests to highly regarded recordings, will reveal acoustical concerns that the acoustician can then plan to resolve. Once observations are corroborated and diagnosis of the issues is made, adding *acoustic treatments* to the room helps resolve sound presentation issues. As described previously under Acoustic Properties and Sound, these treatments possess one or a combination of the *absorptive, reflective,* or *diffusive* acoustical properties.

Speaker Monitor Calibration

It is essential to calibrate the speaker monitors and a midpoint volume setting on your monitor controller to your chosen *sound pressure level* (SPL) level. The accepted standard for most audio engineers is 83dB SPL. I usually listen about 70% of the time at 70–75dB SPL, and the 20% at 80–85dB SPL and 10% at 90dB. This calibration process requires an SPL meter and a *pink noise* generator (for example, the *signal generator* plug-in found in ProTools). It is a straightforward and practical way to check for optimal *stereo imaging* and playback performance from your speakers.

1. Open a *pink noise* generator on the PBDAW and set for the *operating reference level* of –14dBFS.
2. Set the *monitor volume* for a midpoint or standard listening setting and mark with a pencil.
3. With a *SPL meter* set to C-weighting[6] at the listening position (on a stand if needed), first mute the right speaker and adjust the left speaker's power amplifier until the SPL meter reads 83dB. Then repeat the same process with the other speaker. Do this for all sets of speakers if you use more than one pair.

This calibration will allow for your monitor controller to have a standardized volume setting for audio at a playback level of –14dBFS to equal 83dB SPL from the speakers. It will help avoid 'volume creep' on long days, and therefore protect your hearing. Additionally, switching from near-field monitors to mains will be seamless, SPL-wise. Finally, the speakers will present recorded audio with the L/R imaging exactly as the mix engineer intended.

Front/Back Treatment Approaches

There are numerous theories for adding room treatments, but two common approaches utilize *front/back* designs. In order to improve *imaging* and manage any sound-reflection based issues, either the front or back of the room is *live* or treated with *diffusers*, and the other end is *dead* and treated with *absorbers*. In a *live front* approach, the front area where the speakers are has diffusive (*live*) treatments, and the back and sides of the room, are treated with absorbers and bass traps to minimize high-frequency reflections and manage low-frequency diffraction and accumulation.[7] Conversely, in order to create a reflection-free zone between the speakers and the listening position, the front can be made dead with absorbers, and the back and sides can be treated with diffusers to maintain an optimal time-delay gap or 'liveness' to the room.[8]

Power and Grounding

A professional mastering studio is preferably *AC power* and *ground isolated*, and ideally powered by an independent electrical panel feeding 20-amp circuit breakers and outlets. This alleviates any ground loops or hums in the system, and allows the equipment to draw the requisite electrical current for optimal functioning. Note that high-wattage power amplifiers or audio equipment may draw more current to reproduce dynamic intensity or louder passages in the music.

Speakers and Power Amplifiers

The primary consideration with main speakers for mastering is that they reproduce full-frequency audio that the Mastering Engineer can relate to accurately so that the finished project *translates* well to other playback systems. Most main speakers are either *two-way* (a woofer and tweeter with a *crossover network* in the cabinet to separate the two frequency bands), or *three-way* (three separate speaker drivers with two *crossovers* that separate the three frequency bands). The extra driver(s) should add definition and detail in the corresponding frequency range, but there can be issues at the crossover frequencies to listen for. With some audiophile speakers, an additional tweeter is added for a *four-way* speaker. A *subwoofer* (or even two) is often implemented so the lower octave and any sub-harmonic frequencies are adequately represented. Additionally, as I discuss later in this section, speakers are either passive (with separate power amplifiers) or active (amplifiers included in the speaker cabinet). Selecting and testing main speakers is a critical aspect of effective mastering studio setup.

The main stereo speakers are placed as two points of an 8–10-foot sided equilateral triangle with the apex being the listening position. Standard speaker mounting approaches are: soffit-mounted (flush wall-mounted to minimize diffraction of sound and especially to direct low-frequency energy into the room), freestanding, or on stands (Sound Anchors™ can be decoupled with spikes and weighed down with sand or lead pellets to minimize speaker cabinet movement and maximize sound energy projected into the room). Acousticians often seek to decouple the main speakers from any structure in the room or building for that matter. Hence, extreme decoupling maneuvers such as stone speaker stands (completely independent of the building structures, which go directly into the earth), or industrial spring-loaded speaker boxes. In this way, the sound energy emanating from the speaker is not dispersed through the floor or walls, and comes directly through the speaker drivers to the listening position. These common options for installing main speakers into the mastering studio are discussed in greater detail following.

Soffit-Mounted

A soffit is a wall-like structure that is custom built to accommodate your full-range speakers at the front wall. There are several advantages to *soffit-mounted* speakers, and you universally see them in professional recording/mix control rooms. For one, they are solid, minimizing energy transfer and movement of the speaker cabinet when sound is played back. Second, they allow the drivers to be flush with the soffit wall (studio front), meaning there are no reflections from behind the speaker cabinet, and no wall behind the speakers to cause smearing and unwanted sound reflections and frequency nodes or cancellations. In a mastering context, there appears to be a trend toward freestanding speakers. This is because if soffit-mounted speakers are problematic sounding, the soffit structure must be rebuilt to accommodate adjustments.

Freestanding

Freestanding speakers either rest on the floor, or on heavy stands. Stands are often filled with sand and put on spikes to decouple them from the studio and diminish energy loss through vibration. The advantage of *freestanding* speakers is that fine-tuning speaker placement or replacing components is relatively easy. Acoustical treatments notwithstanding, when placing *freestanding* speakers, the distance from the drivers to the back wall, sidewalls, ceiling, floor and front wall are measured and checked so they are not multiples of each other—especially if there are reflective parallel or near-parallel walls. This avoids the acoustical issues discussed earlier in this chapter (standing waves, flutter echo, and comb filtering) at the corresponding frequency wavelengths.

Near-Field Monitors

Near-field monitors are bookshelf speakers with 6″ or 8″ woofers that may have frequency response limitations due to both cabinet and speaker size. They are helpful as a second reference for audio. *Near-fields* tend to roll-off in the low frequencies, which can accentuate or reveal the mid- or high frequencies. They can indicate if robust low frequencies apparent on the main speakers need adjustment for impact on smaller speaker playback. *Near-fields* can also be used to mimic real-world playback scenarios that are not ideal, like an inexpensive home system or a car stereo. The Yamaha NS-10 was a *near-field* recording control room staple for many decades, and many engineers grew accustomed to its sound—slightly edgy and rolled-off in the low frequencies. You may occasionally see them in a mastering application as an additional reference speaker.

Off-Axis Listening (OAL)

This is listening without placing your head at the apex of an equilateral triangle with its two remaining sides indicated by a set of speakers. *OAL* utilizes an ancillary set of *near-field* monitors such as Yamaha NS-10s or single-driver 'speaker cubes' such as Auratones that are on the floor or under the console. This is a more drastic method of mimicking real-world playback scenarios. The mastering studio represents a purpose-designed laboratory of sound—often from the ground up. However, real-world listening is not like this, and often drastically different. End-users can listen back to your labor of audio love on tiny earbuds, as a lossy .mp3 file, in a car, or on randomly placed bookshelf speakers in a kitchen, dorm room, living room, etc. To that end, it remains beneficial to have a set of *off-axis speakers* available so you can check how the master translates in a less-than-ideal playback scenario. You will immediately know if your mid-range is too light, or if your low frequencies are building up. Small adjustments like this revealed by *off-axis speakers* can make for a master that *translates* very effectively to real-world end-user playback settings.

The quality of your speakers is mission critical in mastering. They must reproduce the full range of audible frequencies, and even beyond. Depending on the manufacturer, mastering speakers are engineered to reproduce sound from 20Hz or below at the low end, and up to 20kHz or above at the high end of the audio spectrum. You must accurately relate to what you are hearing, and the audio you adjust must discernibly *translate* to all other environments. PMC, ATC, Tannoy, Bowers & Wilkins, or Lipinski are respected manufacturers of full-range mastering speakers.

Power Amplifiers

Standalone power amplifiers are used to power *passive* speakers. Today, many high-end speakers are *active*, meaning the power amplifier is inside the speaker cabinet, and matched to the drivers in regard to power and impedance specifications. This helps create a consistent sound for the speaker if you happen to move them between studios. *Passive speakers* require a power amplifier, and in mastering applications, it is common to have one power amplifier for each speaker, or *monoblock power amplifiers*. This is to maximize the efficiency of each speaker/amplifier relationship, and allow them to be independently responsive, power-wise, to transients and program information. Additionally, *transient quickness* is reproduced with clarity, absent the sag or flattening characteristic of insufficient amplifier power, especially if both speakers share one stereo power amplifier.

This is where audiophile amplifiers incorporating tube and transformer circuit design with point-to-point wiring have appeal. The sound reproduction is rich, with quick transient response and a vitality that contributes to a compelling *listening experience*. Manley Labs manufactures respected examples of tube *monoblock power amplifiers*, whereas Bryston, Classé, and Hafler are known for solid-state offerings.

Monitor Path/Monitor Control

For mastering and critical listening, always seek the simplest but highest quality path of components and wire between the DA converter and speakers. As discussed in Chapter 1 under Playback System Considerations, each component in the playback chain is like a pane of glass[9] and each pane must be transparent so that the audio you hear remains unadulterated. A high percentage of the audio you will listen to, analyze and master will originate from a digital source file. The playback system example in Figure 1.1 illustrates five 'panes' (eight if you count the wire) between source file and speaker. Strive to avoid long cable runs or extraneous elements that can compromise sound such as patch bays, daisy-chained cables, or additional gain stages in the *monitor path*.

For mastering applications, a monitor controller has the following features: a stepped attenuator (so that there is a separate precision resistor set for each attenuator position ensuring accurate volume attenuation and L/R imaging), L/R mutes, mono button, L/R 180-degree phase inversion, and even an active/passive option. Examples are: Custom Manley (Figure 3.4), Crane Song Avocet (Figure 3.5), and Dangerous Monitor (Figure 3.6).

Figure 3.4 Manley custom monitor controller for mastering. Features include selectable inputs, L/R mute, L/R reverse, mono, VU meters, stepped attenuator with precision resistors, meter and output trims, switch for two sets of monitors, passive or active operation.

Source: (author collection)

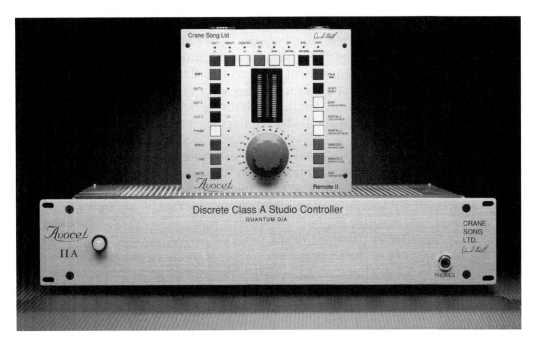

Figure 3.5 Crane Song Avocet Discrete Class A Studio Controller. Features include selectable analog/digital inputs, L/R mute, mono, dBFS Meter, stepped attenuator with precision resistors, stereo or surround speaker selectors, talkback, dim switch.

Source: (courtesy Crane Song)

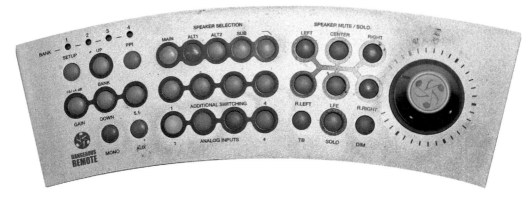

Figure 3.6 Dangerous Monitor Controller Remote. Features include four analog inputs, mono button, stepped attenuator with precision resistors, three speaker selections and a subwoofer selector, 5:1 surround speaker mutes, talkback, dim switch.

Source: (courtesy Capitol Mastering)

Signal (Channel) Path

The *signal path* represents the mastering chain with which the audio is processed, enhanced, or otherwise adjusted. In mastering, analog equipment may be daisy-chained together as needed. However, diligence is necessary to avoid inadvertent loading or frequency roll-offs between analog equipment devices with incompatible circuit designs. This can be tested by sweeping test tones between 20Hz and 20kHz through the mastering chain and listening to/VU metering the end of the audio chain flat (all equipment bypassed), then engaging the equipment (with settings flat) and noting any roll-off in the high or low frequencies at the VU meter. After successfully sweeping 20Hz–20kHz with no issues, play music you are familiar with through the chain and verify that the flat audio matches the post-chain audio by A/B-ing the corresponding positions on the monitor controller.

Another way to set up the mastering *signal path* equipment is to have it on the bypass switches of a mastering console such as the Dangerous Master (and Liaison), Maselec MTC-1X, or a Crookwood Mastering Desk. This is advantageous, as it allows for the quick insertion or removal of an EQ or compressor from your chain to evaluate its effectiveness. Additionally, the buffer amplifiers for each switch alleviate possible loading/impedance issues described in the previous paragraph. The Liaison features three flip buttons to swap the sequence of equipment in the chain, and also a parallel blend fader for parallel processing, which is extremely handy. The selection and sequence of equipment in your chain remains of critical importance, so when a mastering chain sounds special . . . remember to document it!

Mastering Console vs. Point-to-Point Considerations

Some theoretical and practical considerations arise when deciding on a signal path approach. A console adds insert buttons and buffer amplifiers, meaning more electronics and potentially increased noise. Conversely, any loading/impedance issues between analog devices is eliminated. Also, ease of use, quick equipment auditioning, and the ability to swiftly recall mastering chains represent huge upside considerations.

For many years, I bypassed a noisy console in favor of connected equipment point-to-point, and the results were great. In recent years, I have implemented the Dangerous Master/Liaison (Figure 3.7), and appreciate the user interface and the ease in accessing the array of equipment connected to it. With this setup, I only require two DA converters in my signal routing/monitoring setup, because the console features a multed input tap for monitoring of the flat signal. Prior to this, the point-to-point method required three DA converters for a standard mastering setup—one for the flat mix, one for the PBDAW output (mastering chain input), and one for the RDAW output.

Cabling/Interconnects/AC Cords

High-quality cabling and interconnects are advised for mastering applications. Use short runs of high-quality braided copper cables and interconnects: Canare, Mogami, and Van Den Hul are superior examples. These will connect all analog equipment in the channel path via console inserts or point-to-point (*zone 2* of the mastering system), from each DA converter to the monitor controller, and from the monitor controller output to the power amplifiers and speakers. For International Electrotechnical Commission (IEC) AC connectors, I recommend the ESP MusicCord Pro. Research and

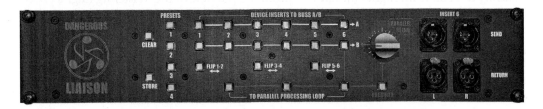

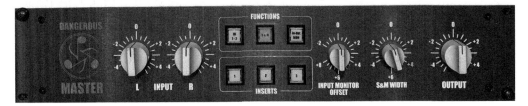

Figure 3.7 Dangerous Master, Liaison, and Convert-2. These three Dangerous Audio devices function together as an effective mastering console. Master features include L/R input balance, input switch for two independent sources, S&M (Side & Mid) functionality for insert #2, in-out monitor switch for flat/mastered A/B-ing, input offset to gain match the flat source to mastered, S&M width, and output gain to drive the next stage of the system. Liaison features include device insert expansion which can be connected to one of the insert switches on the Master (1 or 3 is best, as 2 has S&M functionality), flip options to change sequence of equipment, parallel blend for parallel processing, and easy front panel access on insert 6 for additional equipment. Convert-2 features a high-quality DA converter, clock selector, onboard output calibration, four input selectors, dBFS Meter, external word clock input, and output trim.

Source: (courtesy Capitol Mastering)

audition cables, interconnects, and power cords (IEC connectors), but use common sense and look for cables with a durable build and gold connectors.

Cables and connectors for a mastering studio are designed for routing either balanced or unbalanced analog or digital signals. The most common analog options are balanced XLR, balanced ¼″, unbalanced ¼″, and unbalanced RCA. In the digital domain: AES/EBU, S/PDIF (coaxial RCA), and BNC are the most common configurations.

Digital Clocking and Dither

Choosing between *internal or external clocking* options for each DAW or digital hardware piece in the mastering system remains a matter of preference and debate. Digital audio through each device is clocked at the chosen *sampling frequency* via its clock source (usually internal, external or digital input). *Internal clocks* are found in: I/O cards, AD and DA converters, and any device that performs *digital signal processing* (DSP). Stable digital clocking is critical for the accurate digitization of analog sound at an *AD converter*, or for its reproduction at a *DA converter*.

Two issues can result from an unstable clock—*jitter* and *quantization error*. *Jitter* results from any variation in the sampled signal and causes noise and distortion in the signal. *Quantization error* is distortion characterized by the difference between the actual signal value at time of sample, and the quantization interval value it is rounded to. As signal level decreases, this distortion increases and is rectified by adding a low-level noise known as *dither*.[10] The quality of *DA conversion*, *AD conversion*, and the selection of *dither* for final *rendering* to 16bit from higher bit rates is within the purview of the Mastering Engineer, so take time to research and familiarize yourself with these devices and processes.

There is a subjective philosophy/trend that a dedicated *external clock* can enhance performance and sound quality of DAWs, *AD converters*, *DA converters*, and other digital devices. Hence, the ubiquitous *word clock in* and *word clock out* BNC[11] connectors on most digital audio equipment. These same devices can function as a *master clock*, or an *external master clock* can be connected to clock all the digital devices in the mastering system—known as a 'star' clocking setup—as each clock destination represents a point of the star. Listening tests, A/B comparisons, and the complexity of the digital hardware setup all inform clocking preferences. It remains your choice whether to use a *master clock* for the digital equipment in your mastering system, or to allow the *internal clocks* of your DAWs and/or *AD converter* to clock the system.

The zeitgeist changes; for years, Mastering Engineers favored *external clocking*, but I've noticed in recent years that *internal clocking* has regained favor. It is increasingly accepted that the *internal clock* in an *AD* or *DA converter* performs best, and *external clocking* may provide a different sonic quality but actually reduces clock accuracy and increases *jitter*.[12] I usually clock the PBDAW internally, but my RDAW is clocked by the AES/EBU output from the *AD converter*. I have also achieved excellent results using an *external master clock* for the entire system. This can involve high-quality *sample rate conversion* (SRC) of source audio to match the sampling frequency resolution set by the *AD converter*. The question remains if it improves the fidelity of the final product. Your own research will come into play in deciding how to best implement digital clocking in your mastering system. Prevalent master clocks are the: Antelope Isochrone Trinity (Figure 3.8) and Isochrone 10M, and the Drawmer M Clock Plus.

Figure 3.8 The Antelope Isochrone Trinity Master Clock features an oven-controlled crystal oscillator and sampling frequencies up to 384kHz.

Source: (courtesy Capitol Studios)

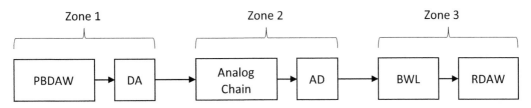

Figure 3.9 Block diagram of The Three Zone Mastering System. Coincides with Figure 3.1, and shows broad organization of the mastering system.

THE ANATOMY OF A PROFESSIONAL MASTERING STUDIO

The Three Zone Mastering System

The Three Zone Mastering System (Figure 3.9) refers to how the signal is routed through the studio for both processing and monitoring. It is also didactic and helps students conceptualize and understand *signal flow* in a mastering context. The *zones* are delineated by the PBDAW *DA converter* and the *AD converter*. In my experience, an ideal *mastering system* utilizes two DAWs to process audio for high fidelity optimization: one DAW for the playback of source audio (PBDAW), and one DAW to record the processed audio (RDAW). This affords the Mastering Engineer a great deal of flexibility in determining not only how, but where in the system to best optimize the audio. Additionally, it allows for audio playback at one *sampling frequency* and audio capture at a different *sampling frequency* without having to SRC the source audio; or conversely, it will support up-sampled source audio so that DSP can occur at a higher *sampling rate* and corresponding *bandwidth*.

My *mastering system* functions in three distinct *zones* for processing audio. *Zone 1* is contained within the PBDAW (digital domain), where level adjustments or preliminary processing via plug-ins before the initial *DA conversion* occur. *Zone 2* is between the *DA* and *AD converters*, and completely analog. At minimum, I use a high-quality analog compressor and an analog parametric equalizer before the audio is routed to the *AD converter*. *Zone 3* is also in the digital domain and includes another instance of DSP, usually a brickwall limiter or digital processor, before the final capture of the mastered audio into the RDAW. Once in the RDAW, there may be some very subtle additional adjustments, but generally, the mastering is complete once the audio is captured as a digital file. It then needs to have the remaining *zone 3* procedure of editing the *tops and tails* and being rendered into the correct formats for delivery and/ or archive.

CONCLUSION

To plan, select, and assemble the components of a mastering studio remains both demanding and exhilarating. If the studio subsequently becomes the source of excellent sounding masters, the Mastering Engineer has reason to celebrate. It is additionally a fantastic exercise—one I assign my students—to become familiar with both the equipment offerings in the pro audio market and the numerous decisions that must be made in setting up a mastering studio. I recommend that the reader conceive, design, and outfit an imaginary mastering studio in order to gain valuable insight into the inner workings and functions of the anatomy of a professional mastering studio.

EXERCISES

1. Select the main equipment for an imaginary mastering studio. Utilize three different budgets: low, middle, and high. Include: speakers, a console, plug-ins, analog equipment, converters, and DAWs. Research and compile a list, and then explain your selections.

2. You are designing a mastering studio in an existing 18′×18′ room. How exactly will you place the speakers and determine the listening position? The floors are linoleum, and the walls are sheet rock. Which acoustical treatments will you add to the room, and why? How will you test the room and speakers for optimal audio fidelity?

NOTES

1. Jeff Cooper (1996) *Building a Recording Studio, 5th Edition,* Los Angeles, CA: Synergy Group. pp. 151–165.
2. Jeff Cooper (1996).
3. Jeff Cooper (1996).
4. F. Alton Everest, and Ken C. Pohlmann (2015) *Master Handbook of Acoustics, 6th Edition,* New York, NY: McGraw-Hill. p. 144.
5. F. Alton Everest, and Ken C. Pohlmann (2015) p. 85. All frequencies from 20Hz–20kHz with equal power per octave so that level decreases as the frequencies rise, emulating human hearing, and is used in acoustical measurements.
6. David Howard, and Jamie Angus (2012) *Acoustics and Psychoacoustics, 3rd Edition,* Amsterdam: Elsevier/Focal Press. p. 85. C-weighting is more suited to sound at higher absolute sound pressure levels, and because of this is more sensitive to low frequency components than A-weighting.
7. Jeff Cooper (1996).

8. F. Alton Everest, and Ken C. Pohlmann (2015) pp. 456–466.

9. Robert Harley (2015) *The Complete Guide to High-End Audio, 5th Edition,* Carlsbad, CA: Acapella. pp. 2–3.

10. Ken C. Pohlmann (2011) *Principles of Digital Audio, 6th Edition,* New York, NY: McGraw-Hill. pp. 36–45.

11. A Bayonet Neill–Concelman (BNC) connector is a miniature quick connector used for coaxial cable.

12. Hugh Robjohns (June 2010) *Sound on Sound: Does Your Studio Need a Digital Master Clock?* Cambridge, United Kingdom: SOS Publications.

CHAPTER 4

Fundamental Mastering Tools and The Primary Colors of Mastering

I've reviewed how an effective mastering studio represents a symbiotic component of the entire mastering system. Next, we will focus on the equipment used to make the tonal or surgical adjustments and sonic enhancements. These include the following five *fundamental tools* for audio adjustments: *analog-to-digital* and *digital-to-analog* converters (AD/DAs), *equalizers* (EQs), *compressors/limiters, expanders, brickwall limiters* (BWLs), and *digital audio workstations* (DAWs). I refer to three items on this list—EQs, *compressors* and BWLs—as The *Primary Colors of Mastering* due to their seminal importance in audio mastering. In this chapter, I will review the function and desirable attributes of these *fundamental mastering tools*.

ANALOG-TO-DIGITAL AND DIGITAL-TO-ANALOG CONVERTERS (AD AND DA)

As their names suggest, these devices convert audio stored in the analog (signal-based) domain to the digital (sample-based) domain, and vice versa.[1] Digital audio must first be converted to analog signal to be heard or used with analog equipment. At minimum, a professional mastering studio requires two or three *DA converters* and one *AD converter*. Two *DA converters* are needed for the PBDAW—one for monitoring flat or unprocessed source audio, and another for the channel path to make adjustments and enhancements to the audio. A third DA is needed to monitor the final processed audio from the RDAW. If you use a modern mastering console that features a *mult* (duplicate) of the input signal, then you only need two *DA converters* for the mastering system, as the PBDAW signal can be accessed after the *mult* for both the flat audio monitor position and the channel path. Additionally, the system requires one *AD converter* to convert the signal from the analog processing chain to digital and then into the RDAW.

Ideally, a mastering-quality *DA converter* will present audio that is focused and accurate without adding coloration, distortion or enhancements. It should be able to convert all common digital audio formats[2] and resolutions to analog transparently. A mastering-quality *AD converter* will also accurately sample and digitize the analog signal into the digital binary code—free from clock jitter and aliasing filter artifacts—at all common sampling frequencies and bit depths. It should preserve transients and dynamic range exactly as they are presented by the analog signal.

Mastering Reference Levels

DA and *AD converters* (Figures 4.1–4.5) usually include onboard level adjustments for calibrating the chosen operating *reference level* for the mastering studio. This calibration correlates digital *decibels full scale* (dBFS) levels with analog voltage readings and the VU meter

Figure 4.1 The Lavry Gold AD122–96MX is an exemplary mastering AD converter. It is sonically transparent and features low jitter, musical clipping properties near 0dBFS, soft limit and soft saturation functions for loudness enhancement, full scale metering, and onboard test tones.

Source: (courtesy Lavry)

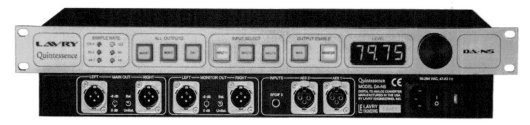

Figure 4.2 The Lavry Quintessence DA converter with onboard monitor control.

Source: (courtesy Lavry)

Figure 4.3 The Lavry Blue 4496 chassis with one AD converter and three DA converters. These converters are also feature-rich and work great in a mastering context.

Source: (author collection)

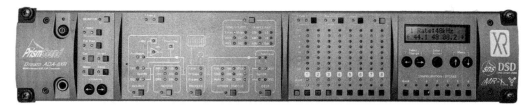

Figure 4.4 The Prism ADA-8XR mastering-grade AD and DA converter with monitoring/headphone functionality. A modular system available in multiple configurations, this unit has the 8AD plus 8DA setup. Also has the 'over-killer' limiting option for creating louder masters.

Source: (courtesy Capitol Mastering)

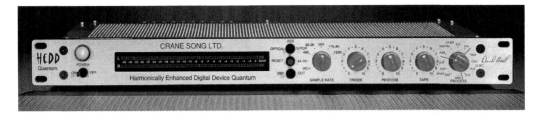

Figure 4.5 The Crane Song HEDD (Harmonically Enhanced Digital Device) features high-quality AD conversion with DSP-modeled analog/tube coloration (triode tube, pentode tube, and analog tape compression options).

Source: (courtesy Crane Song)

Table 4.1 Shows the relationship between a 1kHz tone dBFS level, its output in voltage at the DA converter, the aligned voltage at the DA converter, and the VU meter alignment. Represents an overview of reference level calibration of the mastering studio. The Dangerous Convert-2 has a preset level calibration selector for these three common options (see Figure 3.7).

Mastering Studio Reference Level Calibration			
dBFS (1kHz at RDAW)	**Volts** (at DA/Adjusted DA)	**VU Cal**	**dBu**
−18	.775/1.23	0	+4
−16	.975/1.23	0	+4
−14	1.23/1.23	0	+4

so that lower dBFS levels (adjusted at the DA converter to 1.23V = 0VU = +4dBu) raise quieter more dynamic source files for a more robust level through the mastering system. The *reference level calibration* is accomplished by playing a 1kHz tone at the PBDAW at the selected dBFS level (usually −14dBFS), verification of voltage level at *DA conversion* (1.23V), then to the AD converter, the RDAW and post RDAW *DA converter* for end-of-chain monitoring. *Reference level calibration* is critically important for establishing *headroom* (region of signal between a nominal level and clipping—often 0dBFS digitally and +24dBu analog) in a mastering system, and is usually calibrated as one of three options (Table 4.1).

I have always used −14dBFS = 1.23V = 0VU = +4dBu as my mastering *reference level*. Most mastering studios are similarly referenced, especially when working in modern pop, rock or hip-hop genres.

EQUALIZERS (EQ)

An EQ is one of the three critical *primary colors of mastering*. EQ allows you to boost or cut selected frequencies and bandwidths. As musical instruments possess their own range of frequencies, this tool allows the Mastering Engineer to either feature—or draw attention away from—an instrument or group of instruments. Larger bandwidth or shelving curves are utilized for tonal balancing across larger areas as needed, and narrower bandwidths are useful for detailed or surgical moves. The power of equalization to enhance the fidelity and presentation of recordings remains a profound component of audio mastering.

Equalizer Configurations

There are various types of EQ—*filters, shelving, graphic, fixed parametric* and *parametric* are most of the EQ types you will encounter as hardware devices or plug-ins.

Filtering EQs either remove frequencies above a selected frequency—known as a low-pass filter (LPF), or below a selected frequency—known as a high-pass filter (HPF). The slope of the filter can be adjusted in a *dB/octave* ratio, with 6dB/octave representing the most gradual slope and 24dB/octave the sharpest. The filtering slopes are in multiples of 6dB, with each of four options referred to as an *order*: first order is 6dB/octave, second order is 12dB/octave, third order is 18dB/octave, and fourth order is 24dB/octave. These are excellent for managing high-frequency issues or low-frequency rumbles (necessary for vinyl cutting), and are also useful in conjunction with *shelving EQs* to keep boosts from becoming excessive and causing a tonal imbalance in the high or low frequencies.

A *shelving EQ* boosts or cuts all frequencies above or below a selected or fixed frequency, making it suitable for tonal balancing of highs or lows. *Baxandall EQs* are a shelving EQ curve adopted and mass-marketed as the treble and bass controls on consumer stereo receivers. They have seen a resurgence of interest from Mastering Engineers and professional audio manufacturers alike (i.e. the Dangerous Bax EQ) for a natural and familiar tonality. A *resonant shelf* is a unique shelving curve attributed to Michael Gerzon[3] that dips at the shelving frequency and never plateaus—making it excellent for adding high or low-frequency *extension*, especially in M/S to bring vitality to a lifeless vocal (see Chapter 15—Mid-Side). This type of EQ can be simulated with a wide band on a parametric set for a frequency below 20Hz or above 20kHz, resulting in the same effect, otherwise known as the left or right half of a wide bell curve. A *tilt EQ* will shift along a selected frequency 'fulcrum' and gradually boost or cut to either side of it in a very gradual manner—best applied for tonal shaping.

It is worth mentioning one of the holy grail EQs, the Pultec EQP-1A. The Pultec is a *passive EQ* (using passive electronic components—resistors, capacitors, and inductors—that are not powered until output stage amplification) designed and produced beginning in 1951 by Ollie Summerland and Gene Shank. They can be set for rich low-frequency enhancement and extended shimmer in the high frequencies. There are only two selectable frequency bands and a low-pass filter, but the ability to boost and attenuate simultaneously creates unique and coveted EQ curves that many manufacturers have since endeavored to duplicate. Pairing the EQP-1A with the MEQ5—a three mid-band EQ—allows for more flexibility. Figure 4.6 shows the vintage pair of EQP-1As from Capitol's Studio B.

Graphic EQs have a set series of frequencies at a fixed bandwidth that can be cut or boosted via sliders. These are not common in a mastering context, and are often used for room tuning or live sound (PA) reinforcement applications.

Fixed parametric EQs (example: NTI EQ3 [Figure 4.7] or Maag EQ4M with only HF 'air' shelf selectable) do not allow for the adjustment of bandwidths or frequencies—likely contributing to less expensive manufacture and quick implementation for the user. They have practical applications for mastering, but frequency limitations require using additional EQ options in a mastering system.

Figure 4.6 A pair of Pultec EQP-1A EQs from Pulse Techniques in Englewood, NJ. Pultecs are passive EQs—they don't require power in the EQ circuit and use passive components (resistors, capacitors, and inductors) followed by a tube gain makeup amp.

Source: (courtesy Capitol Studios)

Figure 4.7 An NTI fixed parametric EQ. Stepped attenuators at 3dB clockwise steps, and ¼dB counter-clockwise steps for precise adjustments. 'Air band' is popular in stereo buss/mastering applications.

Source: (courtesy Capitol Studios)

For mastering applications, the *parametric equalizer* remains the primary EQ of choice due to its flexibility in selecting bandwidths (Q)[4] and frequencies. The design and concept is credited to Burgess Macneal and George Massenburg,[5] who designed EQs under the ITI,[6] Sontec, and then later, GML names. This tool allows for precision selection of the independent parameters (bandwidth, boost/cut, and frequency) that will accentuate or attenuate the chosen frequency (or frequencies). To further inform your mastering acumen, it is helpful to familiarize yourself with the following seminal parametric EQs: ITI ME-230 (Figure 4.8), Sontec MEP-250C and MES-432C (Figure 4.9); GML 8200 (Figure 4.10) and 9500; Pultec-style or Enhanced Pultec EQs—Pulse Techniques' EQM-1A3, EQM-1S3, and MEQ-5, Manley Labs' Enhanced

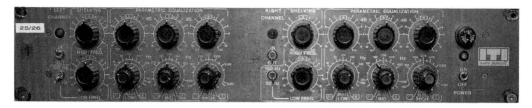

Figure 4.8 The first commercially available parametric EQ—the ITI ME-230 circa 1971.

Source: (courtesy Capitol Studios)

Figure 4.9 Burgess MacNeal started Sontec in 1975 from remnants of defunct ITI, and produced one of the quintessential mastering EQs. Pictured here is one of two Sontec MES-432Cs I use daily at Capitol.

Source: (courtesy Capitol Mastering)

Figure 4.10 George Massenburg's first post-ITI product, the GML 8200 (the GML 9500 is the mastering version with stepped attenuators in 0.5dB steps and precision resistors).

Source: (courtesy Capitol Studios)

Pultec and Mid EQ (Figure 4.11), and the Bettermaker Mastering EQ; Maselec MEA-2; Tube-Tech EQ 1AM; and Weiss EQ1 (Figure 4.12), an acclaimed digital hardware parametric. Notice these mastering-grade parametric EQs generally include the common functionality of a high-pass filter (HPF), low-pass filter (LPF), high shelf (HS), low shelf (LS) and three or four parametric EQ bands for both tonal shaping and surgical EQ approaches. Additionally, they are fitted with stepped attenuators with precision resistors[7] for accurate recall and reliable left/right stereo imaging. Any coloration differences

Figure 4.11 Manley's impressive take on producing mastering-specific Pultecs with stepped attenuators and precision resistors circa 1990–1993. These EQs impart a rich tonality, smooth high-frequency extension and offer the sought-after Pultec EQ curves.

Source: (author collection)

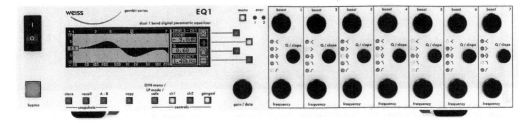

Figure 4.12 The Weiss EQ1 is a seven-band digital hardware parametric EQ. It comes in four configurations (basic, linear phase, dynamic and dynamic-linear phase) and includes M/S functionality.

Source: (courtesy Weiss)

between them are due to discrete components and circuitry design, bandwidth options, the number of EQ bands, frequency options, cut/boost increments, and range.

With the advent of plug-ins, the adjustable settings of a parametric EQ become virtually unlimited, permitting detailed adjustments not feasible in their analog counterparts. These user parameters may include options such as M/S for each band, parallel or series interaction between bands, and phase linear, minimal phase, and analog phase options. Excellent plug-in parametric EQs worth experimenting with are the DMG Equilibrium (Figure 4.13) and the FabFilter Pro-Q3 (Figure 4.14). This precision adjustment of parameters is true for all plug-in audio processing equipment including compressors and limiters—hence my assertion that the best modern mastering system incorporates both digital and analog domains (see Chapter 14—Advanced Mastering Tools).

COMPRESSORS/LIMITERS

These devices manage dynamic range and transient peaks by lowering (compressing) program level above a selected input threshold. They can also be used to add apparent volume or loudness by then adding output gain to the compressed audio. They are used in mastering to keep audio signal from distorting, add sonically pleasing coloration, increase the RMS level of program material (loudness), and create a hyped 'radio' sound. A standard *downward compressor* will compress or lower peaks above a selected input threshold.

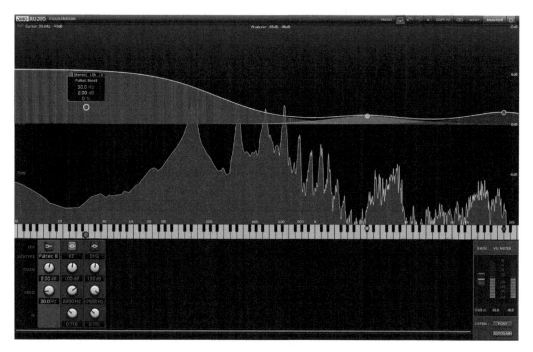

Figure 4.13 The DMG Equilibrium EQ plug-in is feature-rich with multiple filter, shape, phase control, configuration, analysis, and interface display options. It models a hall-of-fame selection of coveted vintage analog EQs (that can be operated linear phase!) including Pultec, SSL E and G series, Sony Oxford, Focusrite ISA 110, API 550, Neve 88, Harrison 32 C, Sontec 250, and GML 8200. Each band can be used in M/S mode. I use it regularly—*the* 'desert island' plug-in EQ.

Source: (courtesy DMG)

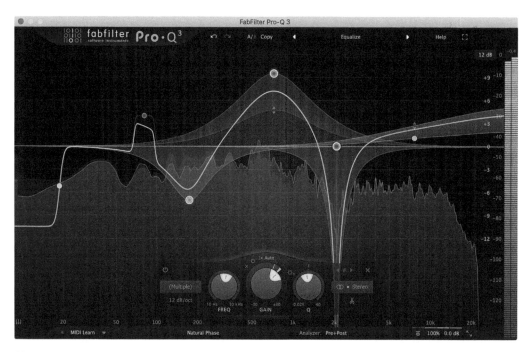

Figure 4.14 FabFilter Pro-Q also boasts a great user interface and impressive features such as phase linear operation, dynamic EQ, M/S processing, and spectrum analysis. The EQuilibrium doesn't have dynamic EQ, but the Pro-Q doesn't have vintage EQ modeling.

Source: (courtesy FabFilter)

However, if you use a compressor in parallel with the main program signal, the combination functions as an *upward compressor*, bringing the 'valleys' up. This is handy for adding detail, clarity, and apparent volume while still preserving transient response, thus maintaining vitality and punch (see Chapter 14—Advanced Mastering Chain Tools and Techniques). Perhaps the purest form of *upward compression* is to raise low sections of a recording with level automation or editing, leaving the transient peaks alone.

Compressors typically have adjustable controls for *input, threshold, ratio, attack time, release time,* and *output*. A brief description of each follows: *input* controls the input level, *threshold* determines the level after which compression will begin, *ratio* determines the amount of compression (i.e. a *ratio* of 4:1 means that for every 4dB of *input* level above the *threshold*, the unit will *output* merely 1dB), *attack* and *release times* refer to the speed of compression action and its subsequent release above the *threshold*, and *output* refers to the output gain (sometimes indicated as makeup gain) added after compression.

By contrast, a *limiter* stops or *limits* transient peaks in program material by using a much higher *ratio* than does a *compressor*. Also, the *attack* and *release* settings are set to more intensely respond to program above the *threshold* setting. For analog *compressors* with *limiting* functionality, such as the Manley Variable-Mu™ or Smart C2, this *limiting* involves selecting from the higher *ratio* settings and adjusting the *threshold* for action on loud peaks, usually with medium to slow *attack* times and fast *release* times. In figurative parlance, a *limiter* is the 'bigger and hairier' version of a *compressor*. A *limiter* can prevent downstream equipment from distorting; or if the engineer raises the overall gain feeding the *limiter*, the average *root mean square* (RMS) level increases, resulting in *louder* sounding masters. For example, the Smart C2 has an 'L' ratio setting for *limiting*, the SSL Stereo Compressor has a ratio setting of 10:1, and the Urei/Universal Audio 1176 has 20:1 as a *limiting* ratio setting. A related note on the 1176—most engineers know of the 'all-buttons' mode that is revered for its renowned 'pumpy' and overdriven sound. I will continue this chapter by primarily discussing hardware *compressor/limiters*, but most of these units have meticulously modeled plug-in counterparts for digital applications.

Varieties of Analog Compressors

Analog compressors are usually designed utilizing one of four different gain reduction-based circuits: electro-optical, field effect transistor (FET), variable-gain, and voltage-controlled amplifier (VCA). A fifth type, a diode-bridge compressor, is less prevalent, but is used in the Neve 2254 and 33609 compressors (see Figures 4.32 and 4.33). Bear in mind that the sound of compressors is also informed by elements beyond the gain-reduction circuit. Transformers or tubes in the circuit design and user-controlled features result in vast sonic differences, so consider all aspects when selecting or auditioning a mastering compressor.

Electro-optical compressors are smooth and musical, and produce warm coloration. They perform gain reduction by virtue of a photocell that reads the brightness of a bulb, light-emitting diode (LED) or electroluminescent panel which is correlated to input level. The classic electro-optical compressor/limiters from the 1960s and 1970s—the Universal Audio Teletronix LA-2A (Figure 4.16),[8] LA-3A (Figure 4.17) and LA-4A—*only* have an input and output adjustment so that *threshold, ratio, attack,* and *release* are internally set. The time constants for *attack* and *release* remained non-linear meaning program dependent response (fast then slow release, for example) and create a slow or 'spongy' response

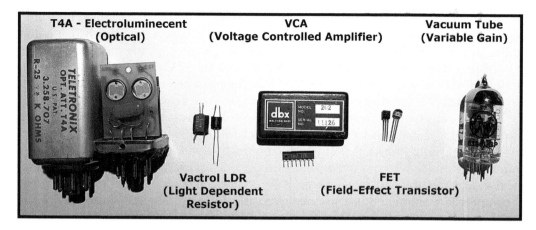

Figure 4.15 Gain reduction circuit component types, from left: the Teletronix LA-2A's renowned T4A optical attenuator with electroluminescent panel and photo resistors, a vactrol light dependent resistor (LDR), voltage-controlled amplifier (VCA), field effect transistor (FET), and vacuum tube.

Source: (courtesy Ian Sefchick)

Figure 4.16 Universal Audio Teletronix LA-2A electro-optical compressor. Originally designed by Jim Lawrence in the early 1950s. Revered for its musical multi-stage release time characteristics, it inspires many modern electro-optical mastering compressor designs.

Source: (courtesy Capitol Studios)

Figure 4.17 Universal Audio/Urie LA-3A electro-optical compressor pair. Uses the same T4 optical attenuator as the LA-2A, but with solid-state electronics (tubes were considered old technology by the late 1960s).

Source: (courtesy Capitol Studios)

characteristic. This makes these compressors well-suited for tonal enhancements and slight peak management. As the demand for stereo buss compression increased, manufacturers added features and functionality to the coveted classic designs to offer stereo electro-optical compressors made specifically for buss compression or mastering such as: the Manley SLAM!, PrismSound Maselec MLA-2, Tube-Tech CL2A, Pendulum OCL-2, and Shadow Hills Mastering Compressor (Figure 4.19).

FET compressors utilize a transistor to perform the gain reduction on the program material. They offer greater versatility and control on transient peaks than an electro-optical compressor since *threshold, ratio, attack,* and *release* are user-adjustable. Characteristic is a focused or intimate type of coloration increasing with extreme gain

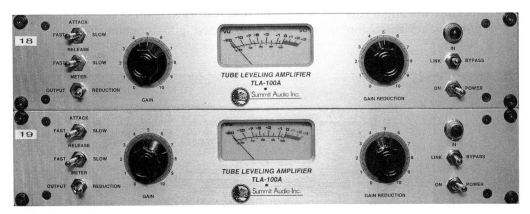

Figure 4.18 A pair of Summit Audio TLA-100A compressors. These are often misconstrued as electro-optical compressors but actually use a proprietary VCA gain reduction circuit.

Source: (courtesy Capitol Studios)

Figure 4.19 The Shadow Hills Mastering Compressor. This unit boasts two compressors in series: an electro-optical compressor first, followed by a discrete VCA compressor. The output transformer is selectable between nickel, iron, and steel for different coloration options.

Source: (courtesy Shadow Hills)

reduction settings. Classic examples are the Urei 1176 (Figure 4.20), 1178, and 2–1176.[9] A FET compressor can be quickly set up to function as a limiter by increasing the *ratio* to the highest setting and then carefully adjusting the *threshold* (generally higher), *attack* and *release* settings for the desired peak reduction response. Modern units suitable for mastering are the Manley SLAM! (Figure 4.21), Crane Song STC-8 (Figure 4.22), and Overstayer 3706 SFE.

Variable-gain compressors utilize the re-biasing of a vacuum tube to handle the gain reduction duties. This topology does not have a user *ratio* control, as the *ratio* increases along with the input amplitude. These compressors are generally known for a warm or smooth coloration characteristic and provide tonality more than the quick control of peaks. Of course, each unit must be carefully listened to for coloration assessment—one of my variable-gain compressors at Capitol imparts a pleasing high-frequency extension along with the expected 'glued-together' component,

Figure 4.20 Universal Audio 1176 compressor pair. Released in 1968, the 1176 implements a FET gain reduction circuit and is known for fast attack and release time settings, and range of tonal characteristics from clean compression at the 4:1 ratio to compelling saturation in 'all-buttons-in' mode.

Source: (courtesy Capitol Studios)

Figure 4.21 Manley SLAM! This is a unique and extremely useful stereo mastering compressor that features both an electro-optical compressor (à la the Teletronix LA-2A) and a FET compressor (à la the Urie 1176). These can be accessed in series, or independently as required. Each compressor has a number of mastering-centric modes. Early models had a digital I/O option that featured mastering quality Anagram Quantum ADDA converters (discontinued in 2009). The output gain has a very even un-hyped sound. It is a clever mastering-centric box and I use it daily.

Source: (courtesy Capitol Mastering)

Figure 4.22 The Crane Song STC-8 is a discrete Class A FET compressor combined with a peak-limiter and an enhancement circuit for introducing analog warmth.

Source: (courtesy Crane Song)

Figure 4.23 The famed Fairchild 670 variable-gain compressor was developed in the early 1950s by Rein Narma for The Fairchild Recording Equipment Corporation. It was used on many Beatles recordings by Sir George Martin, and was a mainstay in vinyl cutting rooms for decades.

Source: (courtesy Capitol Studios)

whereas other versions of the same compressor do not. Attack times are quicker than an electro-optical compressor, but still not at the peak-managing functionality of a FET or VCA compressor.

The most revered and iconic variable-gain compressor is the Fairchild 660 (mono) and 670 (stereo) (Figure 4.23) from the 1950s. The Fairchild has *input, threshold*, six different *time constants* (as attack/release times),[10] and *output* available for user adjustment. They have a hallowed status in the pantheon of compressors due in large part to one Sir George Martin, who favored them on Beatles recordings.[11] They utilized military-quality components and had vertical/lateral (M/S) functionality for lacquer disc cutting applications, so they were often installed in mastering studios before the advent of digital audio and the compact disc. They gained favor among newer companies seeking to create a reliable modern version, with the Manley Variable-Mu™ compressor (Figure 4.25) becoming a mastering studio staple. Other excellent modern examples of this topology are the Undertone Audio Unfairchild (Figure 4.24), Pendulum 6386/ES-8, Magic Death Eye Mastering Compressor (Figure 4.26), Thermionic Culture Phoenix, and Capitol Mastering CM5511 (Figure 4.27).[12]

VCA compressors utilize a voltage-controlled amplifier whose control voltage is derived from the audio input signal itself to effect the gain reduction. Classic examples are the dbx 160 (and permutations) (Figure 4.28) and dbx 165. These versatile compressor designs offer a great degree of user control for attack/release settings, making them excellent for stereo bus and mastering applications. Common VCA compressors suitable for mastering are the Alan Smart C2 (Figure 4.30), Neve 33609 (Figure 4.33), SSL G-Series (Figure 4.29), API 2500 (Figure 4.31), Overstayer 3722 SVC, and Vertigo VSC-3.

Figure 4.24 The UnFairchild is a faithful modern-day recreation by UnderTone Audio (UTA).

Source: (courtesy UTA)

Figure 4.25 The Manley Variable-Mu™ Stereo Compressor/Limiter is a mastering studio staple that draws on the design of the Fairchild 670.

Source: (courtesy Capitol Mastering)

Figure 4.26 The Magic Death Eye Stereo Compressor (designed and hand-built by Ian Sefchick) is a variable-gain compressor that boasts impressive details, such as a special EQ feature in the gain reduction circuit and hand wound transformers.

Source: (courtesy Magic Death Eye)

Figure 4.27 The Capitol CM5511 Stereo Compressor. Only four of these variable-gain compressors were built in 2011 by Ian Sefchick, and are used regularly at Capitol Mastering.

Source: (courtesy Capitol Mastering)

Figure 4.28 A pair of dbx 160 compressors (designed by David Blackmer in the early 1970s) that use a VCA gain reduction circuit.

Source: (courtesy Capitol Studios)

Figure 4.29 The SSL G-Series Stereo Compressor is a famed buss compressor, and was also onboard SSL consoles of the era.

Source: (courtesy Capitol Studios)

Figure 4.30 After working with SSL and designing the G-Series compressor pictured in Figure 4.29, Alan Smart released the C2 Stereo Compressor. This one is from my mastering studio at Capitol, and imparts a variety of useful tonal characteristics, depending on the setting. 'Crush' mode yields over-compression with a mid-range boost, and higher distortion.

Source: (courtesy Capitol Mastering)

Figure 4.31 The API 2500 Stereo Compressor is a mainstay of bus compression.

Source: (courtesy Capitol Studios)

Figure 4.32 The Neve 2254/E (released in 1969) uses a diode-bridge topology for its gain reduction circuit. They were included onboard Neve mix and mastering consoles of the era, and later often removed and rack-mounted with a power supply—as pictured here.

Source: (courtesy Capitol Studios)

Figure 4.33 The Neve 33609 evolved from the 2254 and was released in 1985. It remains a favorite stereo buss compressor among mix and Mastering Engineers. It also uses a diode-bridge topology for gain reduction. This is an early incarnation with no Neve logo or front-plate power switch.

Source: (courtesy Capitol Studios)

EXPANDERS/GATES

Expanders increase dynamic range in audio by either lowering level below the *threshold* (*downward expansion*) or increasing the level above the *threshold* (*upward expansion*). Downward expanders are more common and function like an 'inverted' compressor by lowering the 'valleys' in the program material. *Expanders* are used far less often than *compressors* in mastering, but can be helpful in working with an over-compressed mix. The extreme version of an *expander* is a *gate*. With a *gate*, very little or no audio is heard when the audio signal passes below the threshold. I mention *gates* here for their relationship to *expanders*, but please note they have no practical application in a mastering context.

BRICKWALL LIMITER (BWL)

A *brickwall limiter* (BWL) is a digital look-ahead peak-limiter that allows for a dBFS setting beyond which no audio signal will pass. Look-ahead refers to the main signal being delayed and the side-chain analyzed so that the limiter can process program peaks with ultra-fast *attack/release* times and a *ratio* of infinity:1. The result is extreme level control with no output signal above a user-set ceiling. The settings are usually *threshold* (with auto-gain makeup in some designs) or *input gain, digital output ceiling*, and *release* or *time constant* settings. A BWL is placed in *zone 3* just after the AD converter, and prevents over-levels at the RDAW, allowing for unhindered loudness maximization. The first readily available BWL was the Waves L1 software in the mid-1990s, then in 2000, the Waves L2 hardware (Figure 4.34) was released, quickly becoming ubiquitous in mastering studios. Other early hardware BWL examples are the t.c. electronic M6000 (Figure 4.35) and jünger loudness control devices.

Informed by the intense limiting of radio broadcast compressors, the BWL proved transformative to the mastering world, ushering in a new era of the *loudness wars*. Responsibly and carefully implemented, a BWL remains a valuable and effective tool for a Mastering Engineer to achieve reasonable *target levels*. However, over peak-limiting has very undesirable artifacts including 'hashiness,' inter-sample peak modulation, inverted dynamics (loud sections shrink, and quiet sections overtake), and an uncomfortable *listening experience*. The BWL democratized a critical component of mastering—loudness—but many mix and Mastering Engineers began submitting peak-limited mixes that were afflicted with the artifacts described previously. Indeed, initiatives such as Apple's *Mastered for iTunes* and streaming platforms such as Spotify that post level specifications are a direct response the pitfalls of overly peak-limited music.

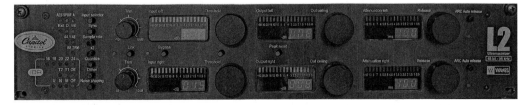

Figure 4.34 The Waves L2 (hardware version) is a digital look-ahead peak-limiter released circa 2000 that evolved from the Waves L1 plug-in. It utilizes 48bit fixed-point DSP and allows for a dBFS ceiling to be set, over which no signal will pass. The L2 was widely embraced and ubiquitous in mastering studios (along with the hardware t.c. electronic M6000 Mastering Processor) and demarcated a significant point in time in loudness enhancement, management, and the fabled loudness wars.

Source: (courtesy Capitol Mastering)

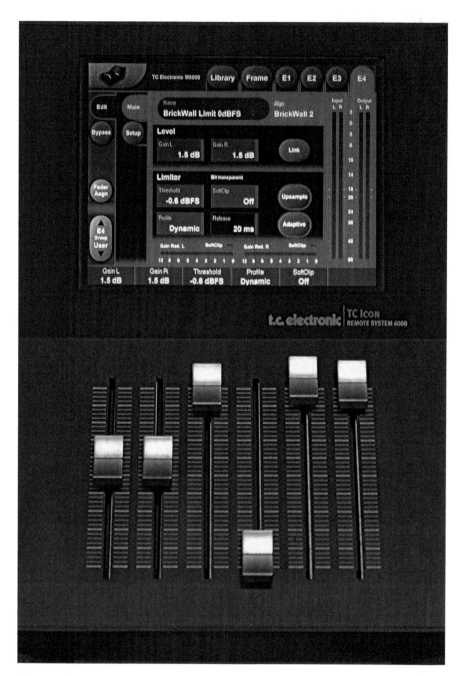

Figure 4.35 The t.c. electronic M6000 Icon Remote, which controls the mainframe M6000 Mastering Processor. The M6000 implements 48bit fixed-point DSP, and in addition to quality peak-limiting, offers vast processing options including EQ, multiband compression, and expansion in both stereo and Mid-Side.

Source: (author collection)

BWL Safety—User Tips

A little peak-limiting goes a long way, and 1–2dB is plenty. One approach I regularly use is to begin with the BWL in bypass, and go for a good genre-appropriate sound and gain structure using the PBDAW and analog chain (*zone 1* and *zone 2*), carefully avoiding over-levels at the RDAW. At this point, you should have a VU level of about +8dBu. Now engage the BWL and make your way to around +10dBu, which may be enough; otherwise, distribute additional gain at appropriate points in the mastering chain (before the AD converter) for a *target level* of around +12dBu. If the BWL is a plug-in, you will capture the relatively dynamic setting of +8dBu at your RDAW, then add a plug-in peak-limiter in the RDAW (*zone 3*). Using a BWL effectively requires judgment, care, and vigilance—the described process may require subtle *macro-dynamic* adjustments at the PBAW to preserve song dynamics. Note that with kick drum/bass-heavy genres such as rap or dance music, the kick will naturally swing beyond the VU meter levels indicated (Figures 4.36–4.38).

Figure 4.36 The Voxengo Elephant is a brickwall limiter plug-in which offers a wide variety of coloration options and customizable parameters.

Source: (courtesy Voxengo)

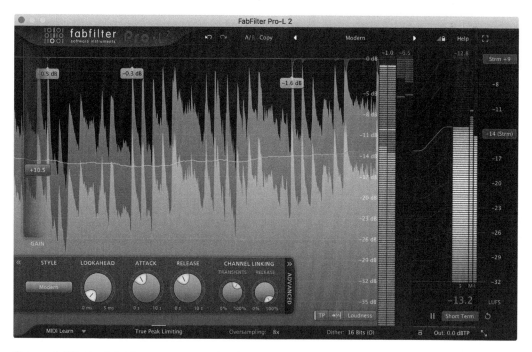

Figure 4.37 The FabFilter Pro-L2 is another feature-rich brickwall limiter plug-in worth auditioning and implementing for mastering. It includes extensive loudness metering.

Source: (courtesy FabFilter)

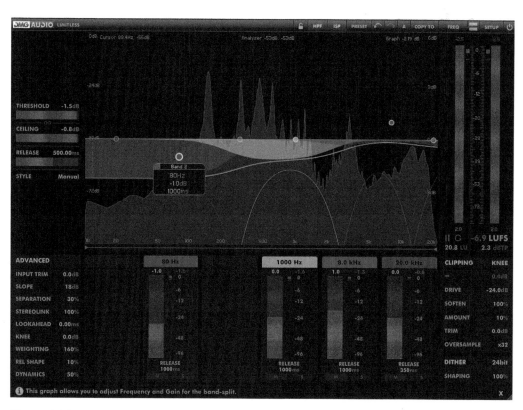

Figure 4.38 The DMG Limitless brickwall limiter plug-in implements multiband dual-stage processing that separates dynamics and transients, and generates extremely smooth gain reduction curves.

Source: (courtesy DMG)

DIGITAL AUDIO WORKSTATIONS (DAWS)

A DAW is the computer-hosted software that allows for the recording, editing, processing, and delivery of the final audio master. For professional mastering, I recommend a two-DAW setup for the following reasons: the playback DAW (PBDAW) remains dedicated to playback, preliminary processing, and source level adjustments; the record DAW (RDAW) only captures the processed audio and renders/generates all required master formats; it allows for different playback and record sampling frequencies/bit depths; and it keeps your files organized as flat and mastered on separate drives. My PBDAW is Avid Pro Tools on a Macintosh, and my RDAW is Steinberg WaveLab on a PC. Some Mastering Engineers are devoted to Macintosh computers, others to PCs, or a combination of the two. I have also experimented with a two-PC system, and for that I recommend the excellent PC Audio Labs Rok Box custom PCs configured for audio work. The most common mastering-specific DAWs are Steinberg WaveLab, SADiE, Sonic Soundblade, Magix Sequoia, and Pyramix.

I realize many engineers use a single-DAW setup, especially those newly developing their mastering approaches. In a single-DAW setup, however, usually both playback and capture functions occur in Pro Tools, which is recording/mix rather than mastering-dedicated software. Additionally, playback and record sampling frequencies must be the same, often requiring *sample rate conversion* (SRC) of the source audio to meet any high-resolution requests. Also, final masters must be assembled in other software that can create a DDP or PMCD Master. I acknowledge it as a viable option, but don't prefer it other than for an in-the-box mastering approach (see Chapter 16—In-the-Box Considerations).

CONCLUSION

Proficiency with the fundamental tools described in this chapter provides a solid foundation of skills for a Mastering Engineer. Well-modeled plug-ins represent a simulacrum of the coveted and expensive hardware equipment described in this chapter at a fraction of the cost, and often with expanded functionality. Despite this, desirable aspects of analog processing are hard to replicate in the computer. Conduct your own research and testing regarding the usefulness and fidelity of ADs, DAs, EQs, compressors, and BWLs. Perform A/B listening tests by changing just one piece in the mastering chain and carefully compare. This will help develop your opinion about what sounds better and why. Anyone can buy audio equipment, but informed and accurate knowledge of the tools help define a great Mastering Engineer.

EXERCISES

1. What three devices make up The Primary Colors of Mastering? What specific function does each device provide the Mastering Engineer?

2. If an all-*analog* signal path is selected, what two additional devices are required in the mastering chain (assuming a digital source file)?

3. Define each measurement represented in the relationship equation: $-14\text{dBFS} = 1.23\text{V} = 0\text{VU} = +4\text{dBu}$. Which measurements are digital, and which are analog?

NOTES

1. Analog audio is defined by a signal or wave; digital audio approximates analog via quantized samples at a fixed sampling frequency.
2. Pulse code modulation (PCM) or Delta Sigma binary codes.
3. Michael Anthony Gerzon (1945–1996) is probably best known for his work on Ambisonics and for his work on digital audio. He proposed the resonant shelf EQ shape described—rising shelf with a parametric dip at the shelving frequency.
4. Bandwidth is represented by the letter 'Q' as a single number ratio. The formula is Q = center frequency/bandwidth (cf/bw). It can also be represented in dB/octave.
5. George Massenburg and Burgess Macneal authored a technical paper entitled "Parametric Equalization" which was presented at the 42nd convention of the Audio Engineering Society in 1972. Macneal would design and manufacture the seminal Sontec MEP-250C parametric EQ in 1975. In 1982, Massenburg founded George Massenburg Labs—among GML's most venerable products are the GML8200 Parametric Equalizer and the GML8900 Dynamic Range Controller, which reacts to loudness like our ears do, rather than to voltage levels.
6. Designed by Burgess Macneal and George Massenburg, the ITI ME-230 (Mastering Equalizer 230) was the first commercially available parametric EQ.
7. Utilizing an accurate stepped attenuator controlled by a rotary switch (i.e. Grayhill, Inc.) and a separate precision resistor value for each position of boost/cut, frequency, and bandwidth.
8. Lynn Fuston (2013) *A History of the Teletronix LA-2A Leveling Amplifier*. See www.uaudio.com/blog/la-2a-analog-obsession/
9. Lynn Fuston (2000) *UA'S Classic 1176 Compressor—A History*. See www.uaudio.com/blog/analog-obsession-1176-history/
10. Hannes Bieger (May 2016) *The Fairchild 660 & 670—Sound on Sound*. SOS Publications.

   All positions offer extremely fast attack values, between 0.2 and 0.8 milliseconds. This was considered to be lightning fast at the time, and the figures weren't beaten until a decade later, when solid-state units with FET gain cells (like the Universal Audio 1176LN) brought attack values down to a mere 20 microseconds. Given its primary purpose as a protective device in broadcast or vinyl cutting environments, in which the limiter was required to catch any unwanted signal peaks reliably, these fast attack times were one of the Fairchild's most important features.

11. Hannes Bieger (May 2016).

   Since the "A Hard Day's Night" sessions in 1964, almost all the Beatles' vocals were sent through the Fairchilds, and the American limiters also played a huge role in shaping the sound of Ringo's drums, the guitars, and many more sources. The Beatles pedigree greatly helped to establish the Fairchild as a classic choice in music-production facilities all over the world.

12. Only four of these were made in 2011, one for each mastering studio at Capitol, by then technician Ian Sefchick.

Professional Mastering Process

The Five Step Mastering Process

In Part I and Part II, I covered foundational aspects of audio mastering, including: the prominent competencies of an excellent Mastering Engineer, the mastering studio, and fundamental mastering tools. In Part III, I present my repeatable mastering system culled from thousands of mastering sessions—The Five Step Mastering Process—that I use daily to create high-quality master recordings.

CHAPTER 5

Step I

Objective Assessment

When you receive digital files for mastering, you must first verify and notate seven *objective assessment* parameters that reveal critical aspects of the audio file. This preliminary data will begin to inform your game plan or strategy to effectively master the project.

In this era of music production, the vast majority of mix files a Mastering Engineer receives are digital files. The stark reality remains that today, the number of clients who provide me with analog master tapes is negligible. And, although it represents an excellent format—with the appealing sound coloration of tape compression—the added cost and time have made mixing to tape nearly obsolete. Whereas *objective assessment* should also be done for an analog source via analog VU/peak meters, for the reasons just mentioned, I will focus on digital source files.

VERIFY THE SAMPLING FREQUENCY (S), BIT DEPTH (BD), AND FILE TYPE

Never take the client's word about the sampling frequency (S) and bit depth (BD) of a source file. It remains your responsibility to verify all aspects of the digital source. A working knowledge of *sampling theorem* (see definition in Chapter 1 under Key Understandings to Possess) remains essential to understand the implications of the digital portion of *objective assessment*. A Mastering Engineer must understand three critical parameters that originate from *sampling theorem* and the information they indicate: *sampling frequency* pertains to the *bandwidth* of the audio; *bit depth* pertains to the *dynamic range* of the audio; and multiplying the S, BD and the number of channels of audio yields the *bit rate* in *kilobits-per-second* (kbps).[1] The *Nyquist frequency*[2] (S/2) is one-half the S and determines the audio bandwidth of the file. For instance, a file with a S of 44.1kHz has a *Nyquist frequency* of 22.05kHz—meaning there will be no audio sampled and reproduced above 22.05kHz.

Use the onboard tools or third-party plug-ins in your DAW to verify these properties of the digital audio file. For instance: in Pro Tools (Figures 5.1 and 5.2), *window* menu then *workspace* displays S and BD; on a PC *right click* then select

Figure 5.1 In ProTools, select Window→Workspace from the menu.

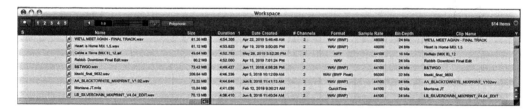

Figure 5.2 Workspace in ProTools will display pertinent information about each displayed file including format, sampling frequency and bit depth.

properties to display the *bit rate* from which the S and BD can be extrapolated using simple arithmetic. WaveLab (Figure 5.3) has an onboard bit meter to verify *bit depth* (or use a bit meter plug-in such as Bitter [Figure 5.4]), a spectrum analyzer to verify frequency response (and thus the *Nyquist frequency*), and loudness meters to verify level. Or, open the .wav file in WaveLab (Figure 5.5), and the S and BD will display in the lower right corner of the main display. Use these tools to verify the parameters/ formats of flat mix files, raw mastered files or files rendered for final delivery. Finally, notate the S, BD, and the file type provided to compile three of the seven parameters of *objective assessment*.

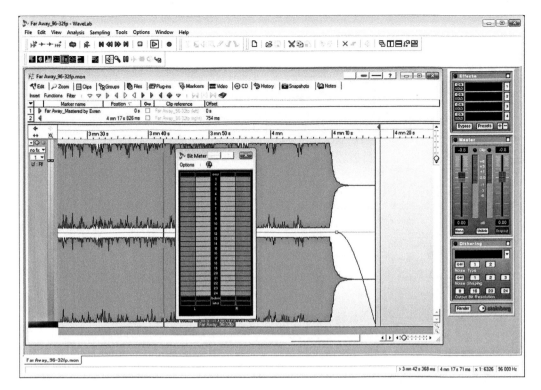

Figure 5.3 Opening the bit meter in Steinberg WaveLab and playing the file will display the bit depth. The audio in this example is 24bit.

Figure 5.4 The Bitter plug-in (freeware) will display the sampling frequency and bit depth as a file plays. This example shows 96kHz—24bit.

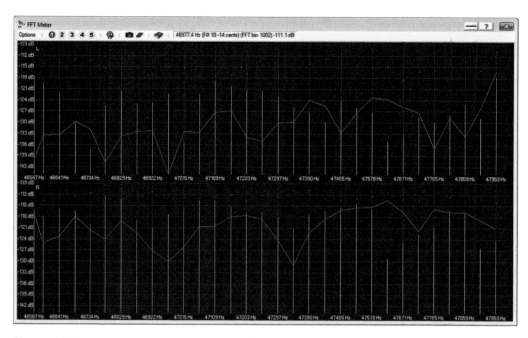

Figure 5.5 The Spectrum Analyzer in WaveLab can be zoomed-in on to verify the bandwidth of a file. This is a 96kHz file since the bandwidth is 47,953Hz (essentially 48kHz), N = S/2.

Figure 5.6 In ProTools, the Gain plug-in (AudioSuite → Gain) will display the dBFS Peak and RMS measurements of a highlighted file.

MEASURE THE DECIBELS FULL SCALE (DBFS) RMS AND PEAK LEVELS OF THE AUDIO FILE

These two criteria represent the fourth and fifth of the seven *objective assessment* parameters. All DAWs have onboard tools whereby you can take these measurements. In Pro Tools, under the *Audiosuite* menu, the *Gain* plug-in has an analyze button that will measure and display these readings for a highlighted file (Figure 5.6). In WaveLab, under the *Analysis* menu item, there is a global analysis (keystroke 'y') dialogue box to check the readings of the mastered file (Figure 5.7). Notate these readings, along with the S and BD. Additionally there are a number of effective metering plug-ins to choose from to obtain this information: Voxengo SPAN (Figure 5.8), Brainworx bx_meter, HOFA 4U Goniometer & Korrelator, iZotope Insight, Waves Dorrough, and FabFilter Pro-L 2 are all good options.

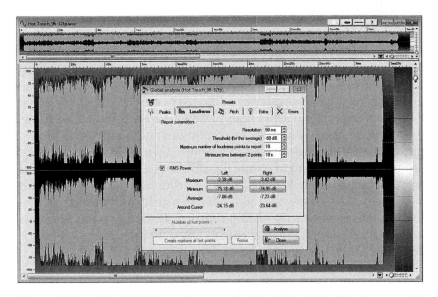

Figure 5.7 In WaveLab, the analysis window (quick key 'y') will measure and display the dBFS Peak and RMS (displayed here) of a highlighted file.

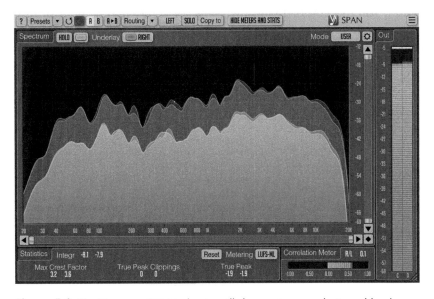

Figure 5.8 The Voxengo SPAN plug-in will do spectrum analysis and loudness metering.

Source: (courtesy Voxengo)

These readings will provide immediate information about how the song was mixed, if the mixer used a peak-limiter, and if it is dynamic enough for an effective mastered result (or if it is already too peak-limited with problematic artifacts). If it means a better sounding final master, don't shy away from requesting more dynamic, less peak-limited mix files to work from.

VU METER READING

Whereas the five digital *objective assessment* parameters discussed previously (S, BD, dBFS RMS, dBFS Peak, and file type) can be ascertained via computer algorithms—the sixth parameter, an analog VU meter reading—must occur in real time. Analog VU meter levels (indicated in dBu) remain excellent for assessing level in audio recordings, particularly in mastering. A VU meter is an analog device calibrated to measure electronic voltage levels, and the ballistics of the meter needles react more to average levels of the audio, so they measure the *root mean square* (RMS) of the continuous audio signal. VU meter ballistics closely approximate the response of human hearing, and remain useful for verifying level cohesion from song to song, and to ensure the album won't be too loud for downstream applications. Carefully listen to the flat mix for fidelity, and also pay close attention to your VU meters as it plays. Notate the loudest levels, usually in choruses or large endings. If, for example, these sections read +6dBu, this is your starting point for mastering—it provides an exact indication of the total gain required through your mastering system to achieve the desired *target level*. See Chapter 1, Figure 1.6, for an example of VU meters.

Target Levels of Mastered Audio

The various music genres possess specific *target levels* once mastered. These represent unofficial guidelines that I've compiled over many years of mastering a wide variety of musical genres. You can conduct your own *target level* research by performing *objective assessment* on your favorite (or charting) recordings.

VU meters provide practical information about the average loudness of both the source mix and the mastered audio. Using VU meters on both source audio and the mastered result allows for a quick real-time calculation of the gain added through the mastering system. Achieving the desired *target level* for the mastered audio represents a crucial aspect of the mastering session.

Table 5.1 VU meter target levels of modern mastered audio by genre.

Target Levels of Mastered Audio:

Genre	Target Level (VU)
Rock (and most sub-genres)	+12dBu—+13.5dBu
Pop (and most sub-genres)	+13dBu—+14dBu
Singer/Songwriter	+12dBu—+13.5dBu
Hip-Hop/Rap	+14dBu—+15dBu
EDM/Dance	+14dBu—+15dBu
Jazz	+9dBu—+11dBu
Non-Limited High-Resolution Audio (HRA)	+6dBu—+8dBu

WAVEFORM INSPECTION

This represents the seventh and final parameter of *objective assessment*. Upon opening the mix file in your PBDAW, you will notice the *waveform* display. Does it resemble a forest, with many jagged peaks visible? If so, chances remain high that transients are intact and it represents a dynamic mix file. This observation should be supported by the

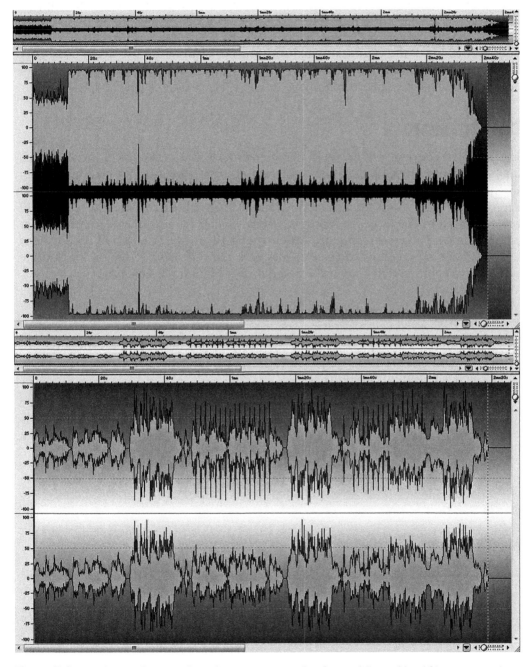

Figure 5.9 Visual Waveform analysis: here is an example of a peak-limited loud file on top and a dynamic file below.

other measured *objective assessment* criteria outlined previously—especially a dBFS Peak less than 0, and an RMS reading below –15dBFS. Zoom into the waveform: Are the peaks rounded rather than flat? If so, this represents additional evidence of a dynamic mix with little or no peak-limiting.

However, if your mix file looks like a block, and when you zoom in the peaks are sometimes sheared off like a butte, then you are looking at a loudness-maximized or peak-limited mix (Figure 5.9). Again, the other *objective assessments* should support this observation—meaning a dBFS Peak of 0 (or between –0.1 and –0.3), and an RMS reading above –15dBFS. If the mix is already at the *target level*, ask the client for a less peak-limited file. If the engineer mixed with the peak-limiter on the stereo buss, have them back down the amount of limiting but leave the device on, as many mix balance decisions were likely made with it on.

CONCLUSION

This completes our study of the seven parameters of *objective assessment*, which establish the fundamental scaffolding of the mastering process. This includes how much gain is required through the mastering system to achieve the chosen *target level*, and what approach to adopt given the degree of dynamic range in the source mix. Next, I will explore Step II of The Five Step Mastering Process, *subjective assessment*. Together, these two assessment processes equip the Mastering Engineer with the data and observations to understand the strengths and weaknesses of the mix file, and decide on the optimal approach to appropriately master the audio for the highest possible fidelity.

EXERCISES

1. Perform *objective assessment* observations on a mix file and notate the seven criteria outlined in this chapter (S, BD, file type, dBFS Peak, dBFS RMS, VU Level, and waveform inspection).

2. Perform *objective assessment* observations on the mastered version of the same file and notate the seven criteria. Compare how the following level parameters changed (dBFS Peak, dBFS RMS, VU meter reading, and waveform display). Write a few sentences explaining the changes and what they indicate.

3. Explain what *sampling frequency* (S) and bit *depth (BD)* reveal about a digital audio file.

NOTES

1. Bruce Fries, and Marty Fries (2005) *Digital Audio Essentials: A Comprehensive Guide to Creating, Recording, Editing, and Sharing Music and Other Audio*, Sebastopol, CA: O'Reilly Media. p. 207.
2. Ken C. Pohlmann (2011) *Principles of Digital Audio*, 6th Edition, New York, NY: McGraw-Hill. pp. 21–25.

CHAPTER 6

Step II

Subjective Assessment

In *subjective assessment*, the Mastering Engineer evaluates the mix solely on the merits of audio fidelity (outlined in Chapter 2—Listening Experience), as opposed to the empirical measurements and verifications of *objective assessment*. It embodies a foundational aspect of audio mastering that constitutes a differentiating factor between Mastering Engineers. If your clients relate favorably to your assessments and the resultant enhancements, it will build your client base and reputation. *Subjective assessment* skills evolve and mature with experience and each completed mastering project. Engaging in music production on any level, where *deliberate listening* is implemented, will contribute to effective *subjective assessment* skills.

DELIBERATE LISTENING REVISITED

I introduced *deliberate listening* (Chapter 1—Competencies) as the substructure of *subjective assessment*. These nested approaches involve listening with a question in mind. Some examples are: What is the genre and/or subgenre of the music? Is it a live-to-two-track audiophile recording, or a multi-track studio recording? Are there layered sounds for richness or impact? What are the discernable studio effects (reverb, delay, double-tracking, phasing, flanging, auto-tuning, filtering, etc.) in the recording? Are these effects enhancing the music or distracting from it? Is the mix dynamic and/or loud? Are there non-musical artifacts or distortion from the recording or mix phase of the project? Are the vocals and diction discernable enough? Are the instruments and/or frequencies balanced appropriately?

Asking and answering these types of questions informs the *subjective assessment* of the audio and codifies the mastering approach. Once you are manipulating equalizers, compressors, limiters, and other audio processing equipment, balancing those adjustments with sensitivity to the music represents a continuation of *deliberate listening* throughout the mastering process.

THREE QUESTIONS THAT FRAME
SUBJECTIVE ASSESSMENT

A few basic questions allow the Mastering Engineer to engage with the *subjective assessment* of source audio. I'll begin with this enduring question from Chapter 1—Competencies:

Is the Listening Experience Pleasing, Effective, and Genre-Appropriate?

While mastering a project, and performing the vast array of processes and adjustments, this question must frame all audio processing decisions. Listen for specific aspects of the mix to assess and contextualize it and ultimately decide: What about the mix is just fine? What is problematic? What could be enhanced or diminished? And how are these auditory observations best incorporated into a *mastering game plan*?

Are There Mix Issues?

This question refers to instrument and frequency balances that require attention. Sometimes a mix has too much or too little low, middle, or high frequencies. Or, possibly a particular instrument (such as the kick drum, guitars, or a vocal) is too prominent or too quiet. Additionally, any distortion or non-musical artifacts in the audio fall into this category. The Mastering Engineer must decide to either solve the issue in mastering or request a revised mix.

How Present or Lacking Are the Eleven Qualities of Superb Audio Fidelity?

The Mastering Engineer must access their audiophile sensibilities in order to effectively answer this question. These Eleven Qualities, described in Chapter 2—Listening Experience, represent key determinants of the quality of a recording, and familiarity with them develops reliable *subjective assessment* skills. Understanding them is critically important, so I relist them here: level, image, transients, depth, definition, clarity, detail, extension, correct dynamics (micro-dynamics, macro-dynamics), vocal presentation, and blossom. Enhancement approaches in mastering are determined by the degree to which these characteristics exist in the recording.

At first, your opinions and assessments may not seem readily accessible, as the requisite depth of creative and technical experience takes time to achieve. However, regular work assessing audio quality will impact these abilities. Indeed, a seasoned Mastering Engineer makes accurate determinations rapidly, and answers these questions after just one or two listens to the recording.

MUSICAL GENRES

Effective *subjective assessment* requires a concrete understanding of the various genres of popular music. Even if you don't naturally gravitate toward certain genres of popular music, you must know what consumers/fans of those genres expect to hear. Regularly listen to various genres of popular music in your studio to keep abreast of trends in the business.

SUBJECTIVE ASSESSMENT BEFORE MASTERING

After *subjective assessment* at this phase of mastering, you either send back the mix for improvements, or accept the mix and move forward to enhancing the mix.

SUBJECTIVE ASSESSMENT AFTER MASTERING

Perform quick *objective* and *subjective assessments* of the mastered file to verify it possesses excellent audio fidelity. If it can be improved, revise or make additional adjustments. Verify that you've attained the desired *target level*, along with any relevant audio characteristic listed in Chapter 2—Listening Experience.

CONCLUSION

The result of *subjective assessment* is a clear picture of the strengths and weaknesses of the flat mix. In concert with *objective assessment, subjective assessment* results in initial approach considerations for the mastering process and a subsequent method for checking the final master. Both assessments inform Step III—The Mastering Game Plan, which is covered in the following chapter.

EXERCISES

1. Complete a one-paragraph written *subjective assessment* of a mix file. Include the genre, musical impressions, and an analysis utilizing The Eleven Qualities of Superb Audio Fidelity listed under Question 3 in this chapter.

2. Assess another assigned mix file and determine if it is suitable for mastering, or if it needs additional mix adjustments first. List the impressions that justify your determination.

CHAPTER 7

Step III

The Mastering Game Plan

After heeding the information and impressions informed by *objective* and *subjective assessments*, Step III is creating and executing a *game plan* or approach for mastering the song or project. This includes selecting: the *sampling frequency* and *bit depth* for mastered audio in the RDAW, the mastering equipment and its sequence in the *signal path* to achieve the desired sonic result, and beginning with *the best sounding mix* for the project. Additionally, The Mastering Game Plan follows four sequential work priorities: 1) Gain Structure, 2) Compression/Limiting, 3) EQ, and 4) Capture. Once this approach is firmly grasped, advanced mastering techniques may be integrated into the sequence. Each mix in the project is approached via these priorities, and you will cascade through them until you are satisfied with the sonic result before moving on to the next mix. I will now explore this and other related aspects of The Mastering Game Plan.

SETUP—THE MASTERING GAME PLAN 'FLIGHT CHECK'

Select Sampling Frequency (S) and Bit Depth (BD) of Mastered Audio

Decide on the *sampling frequency* and *bit depth* for the mastered audio. I prefer 96kHz and 32bit floating-point for the raw mastered file that will be used to render all subsequent formats. It is a great storage/archive resolution for audio, as 96kHz provides a *Nyquist frequency* of 48kHz, and 32bit is an effective storage format for 24bit audio, which allows for 144dB of dynamic range (6dB per bit).

Select Signal Path

At this juncture, the Mastering Engineer selects the equipment and fine-tunes or makes adjustments to the *signal path*. I list analog signal paths here that would be in *zone 2* of the mastering system; however, these same *signal paths* could be created with plug-ins or digital hardware. There are many possibilities for a mastering signal path, but a handful of common ones exist that I will list here. These block diagrams begin with more simple setups, and increase in complexity. Equipment accessed by dashed lines may be present or not, depending on your sonic goals. At either end of each chain would be a DAW (PBDAW on the left or beginning, and RDAW on the right or end):

Path I Example—Compressor First

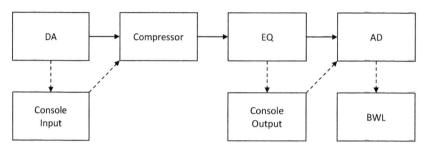

Figure 7.1 Dashed lines show this chain with a mastering console and BWL options, and solid lines show a simpler point-to-point setup.

Path II Example—EQ First

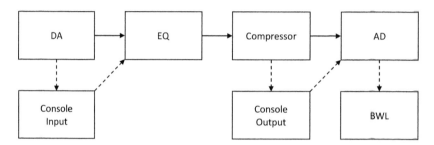

Figure 7.2 Same as Figure 7.1, but with the EQ first.

Path III Example—With EQ First, M/S EQ

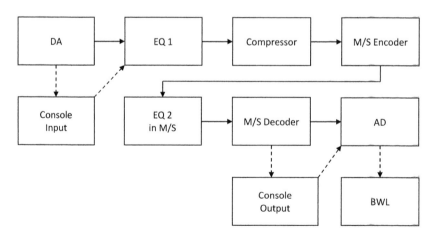

Figure 7.3 Dashed lines again show mastering console and BWL options. Chain is more complex with addition of M/S encoder/decoder and EQ2 in M/S mode.

Path IV Example—With EQ First, Parallel Compressor, M/S EQ

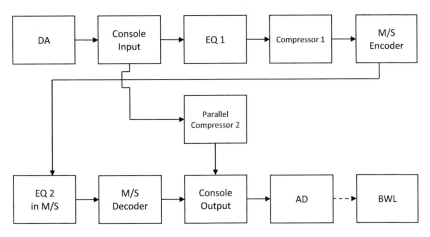

Figure 7.4 This chain makes the mastering console obligatory and adds parallel compression to the chain in Figure 7.3. The BWL is still shown as an option with dashed line.

Path V Example—With Compressor First, Parallel Compressor, M/S EQ, Clipper

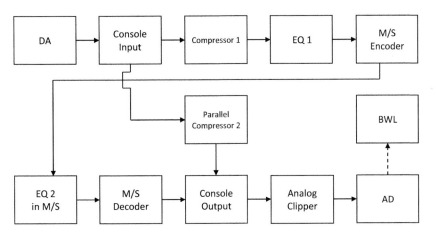

Figure 7.5 This chain also has the mastering console as obligatory and places the compressor first, and also adds an analog clipper to the chain in Figure 7.4. The BWL is still shown as an option with dashed line.

Usually your analog path will include one or two compressors, and one or two equalizers with or without a mastering console. If necessary, this is accompanied by digital signal processing before your DA converter (*zone 1*), after your AD converter (*zone 3*), or both. The most common digital device in a mastering chain is a hardware digital BWL (i.e. Waves L2, t.c. electronic Finalizer, or t.c. electronic M6000) directly after the AD converter, or a plug-in BWL in the RDAW.

AFTER YOU, KIND SIR: COMPRESSOR OR EQUALIZER FIRST IN THE MASTERING CHAIN?

By compressing first then equalizing by adding high frequencies, more shimmer and high-frequency extension can be achieved. By adding low frequencies after compression, low-frequency transient instruments such as the kick drum will swing the VU meters further. More frequency extension is achievable, in other words. This is due to EQ being more apparent after compression. Standard compression will generally smooth out added mid- and high frequencies due to the gain reduction it performs on transients. Note that by adding output gain at the compressor, this darkening is less apparent even making the mix sound brighter (with less dynamic range). Always verify the result is desirable by bypassing then reinserting the compressor or EQ. Take time to experiment with each piece of equipment in order to learn its sonic characteristics. For instance, compressors often have a tonal component that is important to make mental note of; they can sometimes be used as a de facto equalizer in that regard.

By equalizing first, a more compact and contained sound is achievable, but often with more punch, as you can push certain frequencies into the compressor and have it react to them. This configuration works especially well on hip-hop, rap, or R&B music.

MONITORING POSITIONS IN SIGNAL PATH

Monitoring the audio signal at various positions in the mastering chain is critical to effective mastering. At minimum there must be a *gain-matched flat* listening position and a *post-mastering* listening position on the monitor controller. You could also have one at console output (before any post-AD digital signal processing [DSP]), or anywhere you can split signal off to a position on your monitor controller. This allows you to very quickly compare the differences between flat and mastered audio to verify adjustments. There are many options for monitoring, depending on how your mastering console/monitor section is set up.

To avoid over-processing the flat mix in mastering, regularly reference the flat mix while working, otherwise known as A/B-ing. Make sure to honor mix balances while A/B-ing between flat and mastered monitor positions. Your function is to enhance what the artist/engineer/producer created (unless otherwise instructed), rather than completely alter it. The *gain-matched flat* monitor position allows for focus on the EQ and compression changes by removing the more dramatic level change. The goal is to enhance the flat mix in an aesthetically appropriate and pleasing manner; continue the mastering process until this is achieved.

OSCILLATOR SWEEP THE SIGNAL PATH

You must verify that the separate analog pieces in the signal path are compatible with each other. For example, combining balanced devices with unbalanced devices can cause added line noise, frequency roll-off, or signal level loss. Or, equipment with tubes and/or transformers can result in impedance loading issues when paired with solid-state equipment—again resulting in frequency or level loss. The simplest way to check for this is to sweep frequencies through the chain from an oscillator in the PBDAW. First, set the oscillator for 1kHz and monitor the oscillator output at your DA convertor. Make sure it is set at your operating *reference level*. As discussed in Chapter 4—Fundamental Mastering Tools, I use −14dBFS = 0VU = +4dBu (see Table 4.1). Then monitor the same tone after the RDAW DA converter with all equipment in, but set flat—the VU meter should still read 0VU. Next, sweep the oscillator from 20Hz–20kHz and make sure all frequencies give you an even 0VU reading. If there is roll-off, bypass each piece and sweep the tones once again to locate the problem, then either change the sequence of equipment, or use a device with buffered insert switches such as the Dangerous Liaison or other mastering console for better interaction between each piece of equipment. This is a quick overview of a very important lesson—before mastering audio, select, test, experiment, and then implement an effective mastering signal path.

VERIFY L/R INTEGRITY THROUGH MASTERING SYSTEM

Whenever you are patching equipment in and out, there is always the risk of inadvertently swapping left and right, especially if the analog signal path is extensive. Remember, as a Mastering Engineer, all problems are your problems. For this reason, I prefer to set up the Pro Tools PBDAW with multiple mono stereo panels so that I can quickly mute the left or right channel to verify that the corresponding speaker mutes while monitoring the output of my mastering chain at the RDAW output. You could also perform some L/R balancing, but I prefer to do that at the input of the analog console, if necessary.

Select the Critical First Mix to Master

The Best Sounding Mix

Select the best sounding mix fidelity-wise, taking into consideration the recording quality, the engineer, and the studio. If mastering a single, clearly this does not apply. If you are mastering two or more songs as a collection, it remains very important. This is one reason to ask the client how all the songs were recorded and mixed. Many times during the course of making an album, there are multiple studios or engineers working on the project. Some studios/engineers achieve a better sounding result, and you must use that knowledge to your advantage while mastering. In situations where the mixes are similar in fidelity, consider the mix with the most instrumentation, musical complexity, and fastest tempo. This is because it will represent one of the loudest sounding masters in the collection, and the other mixes can be referenced against it.

Organize the Mixes

Group mixes by production team/studio before moving on to the next best sounding production. Ballads or singer/songwriter-type productions can be done by production team group, or left for last. These can remain at a lower volume so as not to overtake up-tempo tracks. Always listen for vocal level cohesion between all songs.

EXECUTION OF THE MASTERING GAME PLAN

Gain Structure

This involves adjusting *gain structure* through the mastering system to reach the *target level* (see Table 5.1). In your mastering chain, there will be several places where you can add gain to achieve your chosen mastered *target level*. The distribution of this gain throughout the signal path is known as *gain structure*, a common term in all aspects of professional audio. For instance, in mixing, it is built up instrument by instrument and monitored via the stereo buss VU meters; in recording, it is the gain from the microphone preamplifier to a compressor to a console fader to the DAW. I get a rough gain setting through the mastering system first, and if re-sequencing of equipment is necessary later, I verify the *gain structure* again.

Let's assume a pop/rock mix has come to you for mastering, meaning your *target level* is +12–+13dBu on a VU meter (see Table 5.1). Make sure your PBDAW has an actual VU meter connected to its output monitor position—or a plug-in VU meter (i.e. PSP VU-3) open on the Pro Tools stereo buss—and set for 0VU = +8dBu. Classic 'textbook' mixes have peaks at around +4dBu VU (the standard alignment of most analog mixing console VU meters); modern mixes can arrive for mastering much hotter. I've found it is ideal to have flat mix playback peaks between +6dBu and +8dBu. If the flat mix *objective assessment* is VU peaks at +4dBu, –3.0dBFS Peaks, and –14dBFS RMS level, 2.5dB of gain can safely be added within the headroom of 3dB at the PBDAW without clipping the DAW software—this playback level change represents your first mastering adjustment. Now you must add a total of 4dB–5dB throughout the remainder of the mastering chain to reach +12dBu–+13dBu. Applying this powerful knowledge frames one of the key functions you will perform as Mastering Engineer—achieving the intended *target level*. The best-sounding option remains a matter of experience, trial and error, and musical judgment. Remember, you can always do A/B tests on short sections of program material to help decide on the best approach fidelity-wise.

GAIN DISTRIBUTION IN THE THREE ZONE MASTERING SYSTEM

Whereas EQs perform frequency-dependent gain, I prefer to conceive of them as tonal devices rather than gain devices. As opposed to analog hardware EQs, many EQ plug-ins include input and output gain adjustments. This can help adjust *gain structure*, especially if there are over-levels at either the input or output of the plug-in, but I prefer to leave these gain settings flat as much as possible.

When mastering, always adjust gain through the *three zones* of the mastering system first (Table 7.1; see also Figure 3.9). Adding gain will change how the audio sounds, and may improve the perception of frequency response, depth, and dimension of the file independent of dynamics or EQ adjustments. Conversely, by EQing first, a frequency band that is enhanced risks becoming over-apparent if gain is added subsequently. This is especially true for mid-range frequencies, where our ears are most sensitive, as the Fletcher–Munson equal loudness contours indicate.[1] Once you have adjusted *gain structure* through the mastering system, and VU peaks are at your *target level*, the next adjustments to tackle are dynamics and then EQ.

Table 7.1 Gain through the mastering system is a crucial first adjustment, best done before compression and EQ.

Places for Gain Distribution in a Mastering System

Location	Mastering System Zone
PBDAW Plug-in (EQ/Compressor/Limiter)	*Zone 1* (before DA converter: digital)
Input of Mastering Console Analog Compressor(s) Output of Mastering Console	*Zone 2* (between DA and AD converters: analog)
Digital Peak-Limiter Post-Mastering Plug-ins	*Zone 3* (after AD converter: digital)

Compression/Limiting (Dynamics)

Once the initial gain structure is set, managing peaks and/or increasing overall apparent volume via compression is important to avoid distortion and over-levels and to add impact to the mix. In a mastering context, compression is generally used to manage and control dynamics in the mix. Access to a variety of good analog compressors remains paramount to a great mastering system. The onboard gain stages often found on compressors are handy for adjusting gain structure to achieve the *target level*. As discussed in Chapter 4—Fundamental Mastering Tools, by adjusting the *input, threshold, ratio, attack, release,* and *output gain* settings on a compressor, an array of dynamics management or level results can be achieved. Proficiency at compression requires consistent practice, but I find that once I discover and like a particular compressor setting, I avoid drastic deviation and use the gain structure through the system to 'hit' the compressor for the desired result.

If more drastic dynamics control is required, a limiter can be utilized. A limiter generally has a higher ratio and quicker release times to manage more intense peaks. Most compressors with 10:1 or higher ratios can be set to function as a limiter. Additionally, in *zone 3* after the AD converter, a BWL is very commonly used to protect the RDAW from any over-levels and also as an option for additional gain (as discussed previously).

EQ

It is best to begin *EQing* after the basic gain structure and *compressor settings* have been adjusted. This is because both gain adjustments and compression can affect the frequency response of the mix, so again, to avoid over-processing, *EQing* is best done third. As addressed in Chapter 4—Fundamental Mastering Tools, EQs boost or cut selected frequencies in the audio spectrum, and even slightly beyond. For mastering purposes a *parametric equalizer* remains most useful. A *parametric equalizer* allows you to independently adjust the frequency, the amount of boost or cut, and the bandwidth around the center frequency. *EQing* is often used for either tonal shaping across broad bandwidths, or it can be used surgically to lift up or diminish specific instruments or frequencies in the mix. Your observations from *subjective assessment* in Step II will inform the EQ approach. Proficiency with EQ is fundamental to the Mastering Engineer and begins with the memorization of frequencies as they correlate to musical instruments and continues with consistent practice (see Table 1.1).

Capture

Once *gain structure, compression (dynamics)*, and *EQ* are adjusted for the desired result (and later advanced techniques), the next step is to capture or record the processed audio into the RDAW. As each track is captured, sample it and previously captured tracks against the next new mix to verify song-to-song cohesion in the collection. Once all tracks are captured, next comes Step IV—Assembly.

CONCLUSION

In this chapter, I've laid out approaching The Mastering Game Plan for a typical session and touched on the tasks that are completed therein. The options are vast for processing the mixes, ranging from digital and analog equipment selection to sequencing in the mastering chain, so experimentation remains beneficial. The setup and execution of approaches that constitute Step III are a vital and important aspect of the *professional mastering process*.

EXERCISES

1. Select a mastering chain without a peak-limiter, utilizing only an EQ and compressor. Take extra care to avoid over-levels. Master three songs and verify cohesion between them. Make sure the VU meter does not exceed a *target level* of +8dBu peaks—about the loudest you can go without utilizing a peak-limiter. Notate flat and mastered levels using the three main *objective assessment* parameters: VU meter level, dBFS RMS, and dBFS Peak.

2. Now, master the same three songs with a peak-limiter added after the AD converter to create. wav files that have a *target level* of +10dBu peaks. Then compare the leading edge transients, frequency response, macro-dynamics, and overall fidelity of the two sets of masters.

NOTE

1. H. Fletcher, and W. A. Munson (1933) "Loudness, Its Definition, Measurement and Calculation." *Journal of the Acoustical Society of America* 5, 82–108.

CHAPTER 8

Step IV

Assembly—Links in a Chain

One day, early in my tenure at Capitol, a respected senior Mastering Engineer at Capitol named Wally Traugott[1] explained to me, "Let me tell you something: if an album is mastered well, you can set your volume and EQ and listen top-to-bottom and you don't need to reach for or adjust the volume or EQ knobs again. The songs flow one to the other like links in a chain." Wally was revered in mastering circles for the prominent artists that sought his work, and Pink Floyd's *Dark Side of the Moon* (1973) was among the albums he had mastered for vinyl release. He would elaborate on that seminal project occasionally: "The producer (Alan Parsons) didn't show up as scheduled, but the tapes had arrived, so I went ahead and mastered it and was concerned that the bell and clock sounds [from the introduction of track #4, "Time"] would blow the cutter head on the Neumann lathe. Alan came in the next day, listened and approved the master." Julio Iglesias, Barbara Streisand, Bob Seger, and other luminaries regularly sought his work, so when he spoke, aspiring Mastering Engineers carefully listened.

Assembly is the process of sequencing and spacing the mastered songs in the correct order for the collection. While mastering a project, a series of adjustments, listening checks, and meter checks are completed to optimize the audio fidelity, level, and loudness of each song as they are *assembled* into a final collection or album. Each song should relate to the others in the album, but be careful: quiet songs, interludes, and production variances (e.g. very sparse or very dense musical information) can be tricky to place properly. By referencing each track (chorus to chorus or verse to verse) throughout the mastering session, audio *cohesion* will be intact and can be double-checked during *assembly*. This final verification of cohesion will bring you a great degree of satisfaction to hear. The rhythm of the mastering session follows a consistent pattern of repeating Steps I, II, and III for each new song in the collection. Song-to-song cohesion represents a fundamental facet of a well-mastered album, which is why the first few songs mastered must be spot-on regarding level, impact, and The Eleven Qualities of Superb Audio Fidelity (see Chapter 2—Listening Experience). There are instances when the sonic perspective of an album takes focus only after completing most of the songs, so never shy away from going back and remastering the first few if necessary.

TOPS/TAILS EDITING AND SONG SPACING

Always edit off the extra dead space at the *tops and tails* of each mastered file. I prefer 300 milliseconds (ms) of silence at the top of each song before audio and an appropriate fade out to silence at the tail. *Song spacing* is to taste but generally ranges from 500 ms to 3 seconds. It can help song transitions to count in time with the tempo of the preceding track and place the new track on a downbeat or upbeat. Include the song spacing after the tail in the file of the preceding track. This allows for an album to be heard as intended if individual files are loaded in sequence (with no additional space added) into a player such as Apple iTunes, or burned to a CD. *Crossfades*—where one song fades out while the next one fades in—should also be edited and fine-tuned during *assembly*.

PQ ENCODING/INDEXING

PQ encoding refers to adding the start index and end index points that determine exactly where a CD player or digital audio player begins and stops playing that particular track. The start and end indexes determine the exact length of the track. It is called *PQ* because this index information is encoded onto subchannels P and Q of the CD. The *PQ information* (along with metadata), is compiled and listed on a *PQ sheet* (Figure 8.1), which is printed and included with a physical CD Master, while a .pdf file of the *PQ sheet* is included in a DDP CD Master folder.

METADATA

Literally 'data about data' in Latin, this is information that is encoded onto the CD Master (DDP Master or PMCD) using your RDAW. It includes *CD Text* and *International Standard Recording Codes* (ISRCs). *CD Text* is usually the song title, and the artist name and will display in automobile or consumer CD Players. ISRCs are distributed to artists and labels via usisrc.org, and each is a 12-digit number that consists of a prefix identifying the label or registrant, the year of creation/publication of the recording, and a series of numbers you chose to identify the specific track. They are usually incremental so that an original album would have an identical but incrementally increased set of codes. ISRCs represent a unique identifier for the song, and help in everything from performance royalty collection, copyright, and verifying the exact version of a recording for administration purposes or downstream releases or compilations.

Please note an obvious exception to *assembly* (which refers to a collection of songs) is if you are mastering a single. In this case, editing the *top and tail* of the file, *PQ encoding*, adding *metadata*, and QC still apply.

CD-ERA SHENANIGANS: HIDDEN TRACKS

In the heyday of the CD era, artists would occasionally include *hidden tracks* at the top of the CD (it would be un-indexed and be placed in the pause of track #1, so that you could only hear it by putting the CD in a player and pressing the rewind button). Alternately, *hidden tracks* at the end of a CD would simply be unlisted on the artwork packaging, but still be indexed (often after a minute, or other amount of time the CD could accommodate). Still other artists would request blank indexes at the end of CD so that the hidden track would play at a specified track index number!

Project : Steve Miller Band "Selections From The Vault [CD]"

Date : Monday, August 19, 2019
Disc Type: Audio
Montage : Steve Miller Band_Selections From The Vault.mon
UPC/EAN Code : 0602577934124
Number of tracks : 13
Time Format : 75 fps
Copyright :

trk	idx	cpy	emp	Title	Time (montage)	Time (CD)	Duration	ISRC
01		x		Industrial Military Complex Hex - Alternate Version				QMEU31908644
	0			Pause	-00:00:02.00	00:00:00.00	00:00:02.00	
	1			Track Start	00:00:00.00	00:00:02.00	00:04:52.72	
						Total	00:04:54.72	
02		x		Macho City - Short Version				QMEU31908675
	1			Track Start	00:04:52.72	00:04:54.72	00:03:23.18	
03		x		Dont Let Nobody Turn You Around - Alternate Version				QMEU31908642
	1			Track Start	00:08:16.15	00:08:18.15	00:04:06.36	
04		x		Love Is Strange				QMEU31908667
	1			Track Start	00:12:22.51	00:12:24.51	00:03:44.31	
05		x		Rockn Me - Alternate Version 1				QMEU31908658
	1			Track Start	00:16:07.07	00:16:09.07	00:03:11.60	
06		x		Crossroads - Live				QMEU31908649
	1			Track Start	00:19:18.67	00:19:20.67	00:07:19.18	
07		x		Take The Money And Run - Live, 2016 Alternate Version				QMEU31908676
	1			Track Start	00:26:38.10	00:26:40.10	00:04:28.73	
08		x		Taint It The Truth				QMEU31908660
	1			Track Start	00:31:07.08	00:31:09.08	00:03:27.31	
09		x		Freight Train Blues				QMEU31908661
	1			Track Start	00:34:34.39	00:34:36.39	00:02:48.49	
10		x		Fly Like An Eagle - Alternate Version				QMEU31908670
	1			Track Start	00:37:23.13	00:37:25.13	00:06:26.34	
11		x		11_I Wanna Be Loved - Live				QMEU31908669
	1			Track Start	00:43:49.47	00:43:51.47	00:05:47.32	
12		x		Jet Airliner - Alternate Version				QMEU31908673
	1			Track Start	00:49:37.04	00:49:39.04	00:04:21.01	
13		x		Lollie Lou - Live				QMEU31908679
	1			Track Start	00:53:58.05	00:54:00.05	00:06:35.37	
				Leadout	01:00:33.42	01:00:35.42		
				Total			01:00:35.42	

1 / 1

---- Created by WaveLab ----

Figure 8.1 A standard PQ sheet that accompanies all CD Masters. Includes start times, end times, and duration of each song and pause, along with all pertinent artist, label, and release information including metadata (universal price listing code [UPC], international standard recording codes [ISRCs] and CD Text).

CONCLUSION

Once the individual songs for an album or collection are mastered, they are *assembled* into the intended sequence. Spacing or cross-fading the songs, *PQ encoding* (indexing), and adding *metadata* all occur during *assembly*. As Step IV of The Five Step Mastering Process, all songs in the collection should be verified to reflect the *target level* of the album. *Assembly* prepares the project for Step V—Delivery.

EXERCISES

1. *Assemble* a three-song project using songs from different genres. Decide on a *target level*, then adjust the three songs so they are cohesive as a collection. Notate the adjustments made to achieve song-to-song cohesion.

2. In the RDAW software, add *metadata* (*PQ Data*, *CD Text*, and *ISRCs*) to the project from the previous exercise.

NOTE

1. Wally Traugott was a Mastering Engineer at Capitol from 1968–1997.

CHAPTER 9

Step V

Delivery—Generate the Master

Once all audio is mastered and approved, tops and tails properly edited, and song spacings set, an *original master* is generated, or final file in the proper format and specifications for replication or archive. The *Red Book Standard* for CD audio—named for the color of the notebook used to document the original specifications—was established by the Philips and Sony corporations to ensure compatibility between player manufacturers in developing CD technology.[1] In 1980, the standard was adopted, and in 1983, the CD revolution was in full swing. A *Red Book* CD Master has a lossless resolution of 44.1kHz—16bit and can be in one of three formats: plant master compact disc (PMCD), disc description protocol (DDP), or digital audio file. An audio mastering DAW can generate masters in all three formats.

PLANT MASTER COMPACT DISC (PMCD)

A PMCD is a *Red Book CD Master*. It contains all mastered/approved audio, and the correct spellings of all titles, artist name(s), and metadata. It includes a *PQ sheet* that is a printout of all the information on the master (see Figure 8.1). The *PQ sheet* lists the client/record label, artist name(s), album name, song titles, and all metadata (ISRC codes and CD Text). It also lists all of the start times, end times, pause lengths, and song lengths. Finally, at the bottom of the sheet, audio quality information or approved artifacts should be listed.

DISC DESCRIPTION PROTOCOL (DDP) CD MASTER

A DDP file represents another *Red Book CD Master* format, and used to replicate physical CDs at the replication facility (plant). It is a folder of digital files that can conveniently be zipped and emailed to a client or replicating facility, or archived on a hard drive. A standard DDP folder (Figure 9.1) contains the following seven separate files.

1. cdtext.bin (CD Text—for display of artist, album and song titles, and timings)
2. Checksum.chk (a value used to verify the integrity of a file or a data transfer)
3. DDPID (identifier)
4. DDPMS (stream descriptor)
5. IMAGE (the audio program as one file)
6. PQDESC (subcode descriptor)
7. .pdf of the *PQ sheet* (information for the P and Q subcodes: CDText, ISRCs, artist name, album titles, and song titles that a CD player will display)

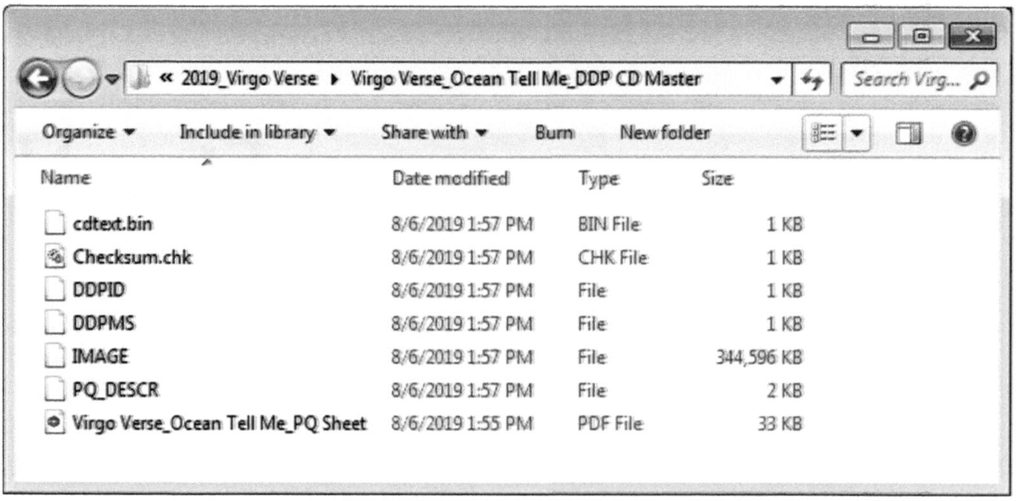

Figure 9.1 Example of a DDP Folder contents.

DIGITAL AUDIO FILES (.wav, .aiff, .dsd, .flac, .alac)

These masters are *digital audio files* that are *lossless* (no data compression). This is most often a .wav (developed by Microsoft) or .aiff (developed by Apple) file at *Red Book* specifications (44.1kHz—16bit), but can also be in a variety of resolutions including 96kHz—24bit, which is common for *Mastered for iTunes* (MFiT)[2] compliant files, and *high-resolution audio* (HRA) files. Digital audio files and streaming platforms are the most recent developments in the distribution and sale of music, profoundly impacting the music industry overall. File sharing had a rocky start with early adopters in college dorm rooms figuring out they could rip files off of CDs and share or duplicate them for free. Labels scrambled to manage and monetize this process that platforms such as Napster made use of. This eventually took place: first Apple iTunes, then Spotify, and then a multitude of other digital music hosting or streaming platforms emerged as the new distribution establishment of the music business, slowly unseating the three-decade reign of CDs. Digital audio files represent a very common master format and can be generated in a variety of *sampling frequencies* and *bit resolutions*.

High-Resolution Digital File Formats (.dsd, .flac, .alac) Overview

These *lossless* formats are used for packaging high-resolution digital files. *Direct Stream Digital* (.dsd) is the file format for *Super Audio CD* (SACD) which was intended to take over the *compact disc* (CD) with higher audio resolution and storage capacity, but it is now popular for audiophile releases. *Free Lossless Audio Codec* (.flac) is another high-resolution audiophile format known for its ability to reduce packaging size of the audio data. *Apple Lossless Audio Codec* (.alac) is Apple's answer to .flac and is supported by iTunes and iOS devices. There is great activity in the high-resolution digital audio market, but no single format has emerged as a standard.

Mastered for iTunes File (MFiT)/Apple Digital Masters (ADM) File

An *MFiT* compliant digital file is a high-resolution file (24bit) that has passed the *MFiT* Codec (available from Apple, Inc. at apple.com), so that it does not contain clipped samples and will not clip or distort when played back on an iOS device. With *MFiT*, the mastering facility converts each song to Apple's specially engineered *lossy* .aac format (using their Pro Plus II codec) and tests it with the *MFiT* user tools codec. Once it passes the codec, the *lossless* high-resolution 24bit .wav or .aiff file is sent to Apple. An *MFiT* file can be any of the standard sampling frequencies (44.1kHz, 48kHz, 88.2kHz, 96kHz), but it must be 24bit resolution. In turn, Apple creates the final *lossy* .aac file that sells on the *MFiT* platform within the iTunes store.

In 2011, Apple released a white paper on *MFiT* describing the ideology behind the initiative and the freeware tools that allow a Mastering Engineer to create *MFiT* compliant files. Each engineer or facility must be officially registered as a *Mastered for iTunes* practitioner. *MFiT* represents a forward-looking approach to optimizing audio for digital file playback and streaming. It also is an unofficial referendum on the *loudness wars*, as overly peak-limited, distorted, exceedingly bright or boomy audio will not pass the codec due to clipped samples. Unfortunately, many music companies understand that the initiative represents a suggestion rather than a strictly policed requirement, and will release *MFiT* designated files that do not actually pass the codec.

Streaming Services

The subscription model of digital audio *streaming services* is now firmly established as a viable platform for music consumers. SiriusXM Internet Radio, Slacker Radio, Spotify, Amazon Music Unlimited, Apple Music, Deezer, Tidal, Google Play Music, iHeartRadio, and Pandora represent most of the popular *streaming services* available today. YouTube and SoundCloud are also very popular platforms for music, and most of these services list audio specifications for submitted content. Whereas it may not be practical to generate a separate file for each platform, a file optimized for streaming or adjusted for level specifications (i.e. Spotify recommends a –14LUFS level) is a consideration for any Mastering Engineer. Additionally, the Sonnox Fraunhofer Pro-Codec plug-in will play a lossless digital file as it would sound in a variety of lossy formats. This represents a beneficial workflow check, allowing the Mastering Engineer to verify that mastered audio holds up in a lossless format.

Files for Vinyl Mastering

Files for vinyl can be to *Red Book* specifications, but I often generate them at 24bit for better audio fidelity for the cutting engineer to use. It is important to bear in mind that the average CD has peaks around +12dBu on a VU meter, whereas the average vinyl LP peaks at around +4dBu. This represents a substantial 8dB difference. The lacquer-cutting engineer will make any final adjustments that are appropriate for the lathe and vinyl format. Even so, I keep vinyl files much more dynamic and quieter overall than files meant for a physical CD or digital distribution. I do this by running an additional pass of each track during the mastering session, implementing reduced limiting and gain so the files do not read more than +8dB on a VU meter.

GENERATING THE MASTER
Dither/Render/Burn

In order to *generate the master*, the raw mastered files must be *rendered* to the chosen final resolution. This almost always involves *dither* if *rendering* from a high-resolution mastered file (32bit floating-point, 32bit, or 24bit) down to 16bit for a *Red Book* master. *Dither* is a type of noise that alleviates quantization error when a file is being rendered to 16bit from a higher bit resolution. *Dither* often includes simultaneous *noise shaping* options that psychoacoustically shift quantization noise away from the audio band or oversample and filter the quantization noise to minimize it.[3] Additionally, 24bit *dither* can be used at the output stage of a DAW with an internal processing resolution of 32bit or 64bit to ensure that there is no truncation of the digital word. The term *burn* refers to making a PMCD in a CD burner; a DDP Master is much more common in this era, and most replicating facilities have a DAW connected to the glass mastering[4] setup so that a PMCD will likely be converted to DDP for the glass mastering process.

Once you've chosen the format, you must *generate the master* in a usable format for replication. And once the actual master is created, you must open it or play it and listen down to the entire completed master to verify its quality and integrity. Next, we'll delve deeper into an important aspect of Step V, *QC*.

Quality Control (QC)

As discussed in Chapter 1, Item 7, QC is the final verification of all aspects of an audio master. It is a critical aspect of mastering, as the mastering session remains the final opportunity to catch and fix problems or make changes to the audio before it is released to the consumer. *QCing* requires listening to each physical master before it is sent out for replication, but it is also done for digital files. It is a formidable responsibility; as a Mastering Engineer you must possess excellent QC skills, along with the vocabulary to describe issues for the artist/label to approve, or to document with the master (often on the *PQ sheet*). Any and all tics, clicks, distortion, indexing anomalies, incorrect metadata, and misspellings remain in the purview of QC. At Capitol Mastering, engineers first spend a few years classified as a Production Engineer whose duties are QC, assembly, and some mastering. It represents a great foundation for the next classification, which is Mastering Engineer. When effectively *QCing* a master, there are beneficial practices to follow that I list here:

Check All Song Starts and Transitions

Quality issues often lurk at the tops or tails of song files or in transitions between songs on an album. Verify audio is not clipped, or ending abruptly.

Listen to the Entire Album

Make sure the song titles match the songs, notate any issues or artifacts (distortion, clicks, tics, pops, drop-outs, etc.) and have the client or label approve them. If approved, notate them at the end of the *PQ sheet*.

Information to Verify

1. Artist, title, selection number
2. Song titles
3. Song indexing: start/end times, including any offsets
4. Metadata: International Standard Recording Codes (ISRCs) and CD Text
5. The *PQ sheet* (named for the P and Q CD channels where the information is stored) which lists: the record label, selection number, artist name, album name, all song titles, ISRCs, CD Text, start/stop index times and offsets, and any quality control notes.

If there is any question regarding the accuracy of this information, you must stop and get approval from the appropriate party, or ask for corrected information.

Audio Integrity to Approve

1. No distortion of any kind
2. No digital over-levels
3. No artifacts or noise, which are not audio or not intended to be a part of the recording
4. The song versions must be correct (clean, explicit, remixed, etc.)
5. There must be level and frequency cohesion throughout each song on the album

My approach to CD Master QCing is to print the *PQ sheet*, verify all data, and then check off each song after it plays successfully. If any anomalies (i.e. distortion, drop-outs, artifacts, level or EQ issues, etc.) in the audio exist that are approved by the label, artist, and production team, I notate this on the *PQ sheet* so that anyone who opens the master and listens down to it can find a documented explanation for any irregularities or problems.

For Digital File Masters

1. Verify: correct file format type, *sampling frequency* and *bit depth*
2. Verify *MFiT* compliance, if so designated
3. Include 300ms of silence at the top (so that a player doesn't clip the beginning), and any album spacing or silence included at the end of the file (so digital delivery results in the same song spacing as a physical album)
4. Measure the length of both the *source file* and the *mastered file* (if using a two-DAW analog mastering chain system); they should match. If not, stop! This indicates a clocking issue at the PBDAW that must be fixed and the song re-loaded.

We've reviewed that a *Red Book Standard digital* file is a 44.1kHz—16bit .wav, and *a high-resolution* (HRA) digital file is 24bit, sometimes with a higher sampling frequency than 44.1kHz (48kHz, 88.2kHz, 96kHz, 176.4kHz, or 192kHz). These parameters must be verified as accurate during digital file QC.

QCing develops the *watchmaker precision and meticulousness* that Mastering Engineers are known for. If you catch a bad edit, misspelled CD Text, distortion, or other mistake before 10,000 copies of a CD are pressed, or before a song is on every digital distribution platform, you will be considered a hero to the record label or production team.

CD Master Designations

An *original master* is the final approved version of an album or a project. A *production master* is a duplicate of an *original master* and is meant for a CD replicating facility. An *archive master* is meant for a library or archive as a backup of the *original master*, and a *reference copy* is meant for the client to listen to and approve (usually a CD or .wav file). Ultimately, the audio on all of these masters is the same, with the exceptions being: different generations of duplication, resolutions of lossless digital audio, adjustments made to an *MFiT* or *Spotify* compliant file (or other digital streaming platform file) that has a recommended LU, level or clipped sample specification.

CONCLUSION

Now that the master has passed QC, it is ready to be sent to a replicating facility or uploaded to an online distribution platform for the consumer. Congratulations! We have now entirely reviewed The Five Step Mastering Process. Take time to memorize these steps, as they will serve you in creating consistently high-quality audio masters. As a Mastering Engineer, you must be prepared to correctly generate any and all requested formats of the mastering project.

EXERCISES

1. Generate a DDP CD Master from a collection of five songs you have mastered and assembled. Make sure that the DDP folder contains seven files.

2. QC the resulting master by printing the *PQ sheet*, verifying all metadata, and notating any concerns.

NOTES

1. Ken C. Pohlmann (2011) *Principles of Digital Audio, 6th Edition*, New York, NY: McGraw-Hill. pp. 187–188, 190. "The Red Book specifies both the physical and logical characteristics of a Compact Disc."
2. *Mastered for iTunes* is known as *Apple Digital Masters* (ADM) as of 2019. It is an initiative by Apple, Inc. to create audio files that will be heard as intended by the production team and that are optimized for iOS playback devices. The original technology brief is at https://images.apple.com/itunes/mastered-for-itunes/docs/mastered_for_itunes.pdf
3. Ken C. Pohlmann (2011) pp. 101–102.
4. Ken C. Pohlmann (2011) pp. 214–218. Glass mastering is part of the Compact Disc manufacturing process at a replication facility involving a laser beam recorder that exposes the pre-master data into a photoresist (emulsion) coating on a glass plate. The emulsion is developed and metallized to produce metal stampers for CD manufacturing.

Professional Mastering Process — Macro Considerations

The foundations of professional mastering investigated in the first three parts of this book reveal the processes that move a mastering session from source file assessment to completed final master. Part IV takes a macro look at effective mastering from preliminary dialogue with the client to inform an approach, adopting an effective mindset, and some tips and secrets from the mastering trenches for compelling results despite unforeseen pressures.

CHAPTER 10

Before the Mastering Session

PRELIMINARY DIALOGUE WITH THE CLIENT

Before beginning a mastering project, it's important to discuss with the client (artist, engineer, producer, manager, or label) about their expectations, your mutual understandings of the specific genre, their current impression(s) of the recording and mixes, and some examples of successful projects that are comparable to theirs. This alleviates surprises and clarifies downstream expectations about the fidelity of the project. If they can send you a near-final mix, it is beneficial as you can verify that it is ready for mastering, or address any concerns before the mastering session. As your reputation and discography builds, clients will rely on your input before the mastering session. Although there can be variances, the most common mix issues to be aware of are over-compressed mix files that have problematic artifacts, improperly balanced frequency ranges (from inadequate room/playback acoustics), and improperly balanced vocals or instruments. Fortunately, all of these concerns can be addressed with revision notes for the mixer.

GENRE-SPECIFIC CONSIDERATIONS

Awareness of musical genres was introduced in Chapter 1 as a subset of being an audiophile. Your clients expect that you understand and are familiar with their particular musical genre and production approaches. Your mastering work and ability to discuss concerns with your clients will benefit from this insight. There is no simple way to access a depth of knowledge about each specific genre, along with its common production techniques. Musicianship or musical ability, familiarity with artists and albums, concerts attended, and personal taste in music will affect this understanding. As such, time spent interning or working in a commercial recording studio is priceless for exposure to these production approaches. With understanding and respect for how the production team created the music, you will be able to connect with the client's musical vision and contextualize your mastering approach accurately.

One particular benefit of my tenure at a major label is being tasked to work on a variety of musical styles that comprise the label's catalog. Also, being in a 'music metropolis' like Los Angeles—with its diversity of musical styles and scenes—offers opportunities to work on a wide breadth of music. You may become known as a rock Mastering Engineer, or a hip-hop/rap Mastering Engineer, but it is beneficial to develop confidence mastering various genres of music. Each genre has different mastering characteristics: target playback level, amount of compression/limiting, frequency and instrument balances, and drums and vocal placement.

EXPERIMENT REGULARLY

One of the senior engineers at Capitol when I began was Bob Norberg, who developed a technique of reversing a source file in the computer and playing it backwards through his LA2A compressors to get a compressed sound without losing the leading edge of the transient attacks. Bob was known for mixing and mastering classical music, and a fan of the genre would call him and complain that in her car, quiet passages would get lost. So Bob utilized this reverse-file method to 'lift up' the quiet passages in classical music. This type of creativity can solve issues and successfully affect the final result, so allow time to experiment and test mastering approaches. You may discover methods or combinations of equipment that produce excellent results in the audio.

THE MASTERING SHOOT-OUT

Professional audio mastering is an extremely competitive arena. Mastering Engineers are known to vie for projects, especially those from 'name' artists, record labels, or producers. In this context, it is common to be asked to *shoot-out* a song first before a final decision is made regarding who will do the mastering.

It's up to you how to manage these requests. The most gentile approach is for the client to pay for the song. Short of that, many Mastering Engineers offer a free snippet or even an entire song. This can feel like considerable pressure, so it is critical to initially discuss particulars about the project and the sound the client is after. Take solace in the fact that your name and reputation allow for you to contend in such circumstances. *Shoot-outs* are naturally done before the session, but on rare occasions or with larger budget projects, an album may be sent to two or more Mastering Engineers simultaneously and the production team subsequently chooses its favorite master.

PRICING

Pricing is an often-ignored aspect of music technology education, but exceedingly important to grasp, as budget discussions precede most mastering sessions. If you work with an office or administrative staff, they can oversee this aspect, but still the subject bears addressing. Make sure that pricing is simple but covers all eventualities and delivery requests (known as 'parts'). For example, one good approach is to set a per-song rate and an analogous hourly rate for revisions. Include one free revision, client-revised mixes at half price, alternate mixes (instrumental, TV, a cappella, etc.) at half price, and a separate stem mastering price. Also include final master pricing (DDP, PMCD), or special additional file pricing (MFiT, Spotify). You can always make a package deal, but it is difficult to go back into a 'priced' project without possibly annoying the client.

CONCLUSION

The success of a mastering session begins with client communication. A clear understanding of the production team's vision for the project will streamline the mastering session and your approach. Familiarity with the musical genre, and related project discourse before implementing The Five Step Mastering Process from Part III, will affect the success of the project. By communicating your understanding of the music, and improving the fidelity of the mix, you will engender trust and respect, which lead to referrals and repeat projects.

CHAPTER 11

The Inner Game

Mindset Approaches

In this chapter, I explore a professional Mastering Engineer's focus of thoughts while mastering. I've identified eight areas that shed insight into an effective mindset, which I detail next.

MASTERING IS EITHER EFFECTIVE OR NOT

A completed mastering project is either effective at enhancing the *listening experience*, genre-appropriate in its presentation, and well executed, or it isn't. It either effectively presents The Eleven Qualities of Superb Audio Fidelity introduced in Chapter 2, or it doesn't (barring recording or mix concerns—it's worth revising a project that doesn't thrill the client). I mention this because with experience and more projects successfully mastered, one develops confidence that each new project has been properly mastered. If mastering revisions are requested, you will possess the professionalism to dialogue with the client, understand their impressions, adjust your game plan, and remaster the project to their liking.

PSYCHOACOUSTICS

Psychoacoustics[1] refers to the study of sound perception in humans, and it is important to discuss a few key points. Ultimately, the process of listening to and evaluating audio is *subjective*. Our ears sense and or brains interpret what is pleasing or displeasing about sound. For example, the Fletcher–Munson Equal Loudness Contours (see Figure 1.3) indicate that mid-range frequencies are more apparent to us, so that our *sensitivity* to them in instruments and music is enhanced. As a consequence, exaggerated mid-range frequencies quickly become uncomfortable to hear. Familiarity also plays a role; expecting to hear audio with a specific sonic quality and impact informs *listening experience*. Imagine a dance or hip-hop track rolled-off in the low frequencies, or a rock track with no mid-range definition in the guitars—the first impression would be: "What happened?" You may have experienced disliking the sounds or production qualities of a new song on the radio, but after repeat listens, later realizing you actually *do* like it.

These are examples of *psychoacoustics* at work. Well-produced audio sounds compelling and appealing, and your experiences listening for and identifying that appeal is uniquely personal. An opinion about the quality of mastered audio is informed by all of your *experiences* listening to music, and your *knowledge* about music and recording/music technology. In one sense, with the exclusion of artifacts or distortion, *psychoacoustics* indicates that there is no incorrect mastering, per se, but everything that informs that particular recording—from style/genre, to recording methods, to end use (function), to context of the artist's other releases, and trends in music—*all* combine to inform how we hear, and how we come to make audio mastering decisions. This knowledge is liberating if the ideal mastering adjustment or alteration escapes you; simple enhancements and adjustments or fresh experimentation are often fine approaches to explore while mastering.

LEAD THE LISTENER BY EMBRACING MACRO-DYNAMICS

Remember that the Mastering Engineer is the proverbial 'wizard behind the curtain' implementing final audio adjustments, and as elucidated in Chapter 2, a primary goal is to enhance *listening experience* for the consumer. Consequently, it remains important to identify and embrace the underlying musicality of the recording in your mastering work. We've covered important foundations and approaches to mastering, but remember that the final result must be appropriately musical. A particular example of this is *macro-dynamics*, whereby you want to *lead the listener* through the various sections of the song to the payoff of the chorus or other musical climax. The song and its recording, mixing, and ultimately mastering should all support the dynamic ebbs and flows of the composition. This is clearly where experience as a musician, songwriter, or producer can pay dividends in mastering. Your clients will understand if you posses a sensitivity to their music when they listen to the final master.

WORK HABITS

A professional Mastering Engineer implements good work habits, allowing them to master audio at a high level of fidelity while also documenting what was done for reference and repeatability. For instance, if the client loves the result, the mastering chain can be recalled for their next release. These habits establish the repeatable structure or scaffolding that allows a Mastering Engineer to hit the mark for the client each time. A key system of habits imparted in this book is Part III—The Five Step Mastering Process; practicing it regularly will produce consistent and professional mastering results.

BE METICULOUS

Mastering as a discipline requires a great deal of organization and attention to detail. There is a potentially wide margin of error beyond aesthetic adjustments that may include improper digital clocking, converter calibration, equipment issues or even failure, compatibility issues between analog devices, and generating the final delivery format at the correct specifications. A Mastering Engineer must successfully achieve aesthetic goals, but also be 100% confident that their technical procedures are sound—another important reason to triple-check your mastering system and document each project with session notes. Your closet may be a mess, and you may not be able to find your tax returns from last year, but you *must* have all of your EQ notes, source files, and mastered files archived and retrievable in short order. Many times, I've restored masters for clients years after our session, and they are beside themselves with relief. Allow the band or artist to get drunk with excitement upon completion of their project—as the Mastering Engineer, you will celebrate, too—but your role is *always* the proverbial designated driver en route to audio nirvana.

With time in front of the speakers (where the client is paying for your expertise), and the project on a deadline, your skills will naturally evolve. You will catch inconsistencies/problems in the audio that range from mixes that are too loud, too quiet, excessive, or lacking in certain frequency bands, to improperly balanced instruments or improper use of the audio image, to name a few. Some of these can be adequately fixed in mastering (for example, sections of a vocal performance that are overly sibilant or 'spattery' can be effectively de-essed without sending the mix back), and others warrant a call to the producer or mix engineer to send a revised mix. Again, don't be shy about requesting revisions, as they will allow you to create a better sounding master, and everyone will be pleased in the end. If the production team wants to provide you with additional flexibility, ask for stem sources.

DOCUMENTATION (NOTES)

I introduced this in Chapter 1 under Competency #9—Implement Recallable Workflow Approaches. You must have an effective system for documenting each project. As mentioned, at Capitol, I repurpose analog tape box legends to take notes, but you can make your own note sheets in Excel, or implement a digital option. Notes should contain session information (date, artist, album title, project title), *objective assessments*, all of the *zone 2* analog EQ, compressor and limiter settings, the sequence of the mastering chain, and also any advanced techniques implemented such as parallel EQ, *zone 1* plug-ins before converting to analog, or *zone 3* processing after capture. This also allows you to track the artists and projects you master throughout the year. An ancillary approach is to keep a notebook handy to jot down workflow ideas, questions, equipment wish lists, client conversations, or other relevant concepts to research later. While working on this book, during mastering sessions, I regularly wrote down important ideas to explore.

LISTEN UNIVERSALLY

Use your ears to listen to *both* the client and the music. Discourse with the client until you are certain you are describing the same musical phenomena with the same adjectives correctly defined. I once had a client who asked for a dynamic master, and then he played me an example of a very limited, loud master. Thankfully, we were able to resolve that discrepancy before I launched into the mastering of the project, avoiding wasted time and unnecessary frustration. Once you and your client are communicating effectively, the mastering session should progress with ease.

MANAGE PROBLEMS WITH HUMILITY AND PROFESSIONALISM

If there is a problem (and there occasionally will be), or the client is not pleased with the mastering, get more clarity on their expectations and redo it. I typically include one revision for free and rarely dispute what the client is hearing before sending the revision (unless we are not hearing/describing the same issues). This is simple professionalism. If the client appears to abuse your goodwill here, you must let them know from the outset that subsequent revisions are at your hourly rate.

CONCLUSION

If your mastering mindset and work habits remain quality-focused and consistent, you will enhance the audio fidelity of each project and continue evolving as a Mastering Engineer. The mindset concepts outlined in this chapter frame mastering work and client interactions for a successful outcome. By understanding and implementing these approaches, you will provide deeper value to the production team, meaning that your insight and skills will be regularly sought after.

NOTE

1. Ken C. Pohlmann (2011) *Principles of Digital Audio, 6th Edition*, New York, NY: McGraw-Hill. p. 336. Psychoacoustics explains the subjective response to everything we hear. (It) seeks to reconcile acoustic stimuli . . . with the physiological and psychological responses evoked by them.

CHAPTER 12

Session-Saving Mastering Tips

In this chapter, I've culled five helpful tips from my daily mastering work at Capitol Records. These concepts are the result of completing deadline-driven mastering sessions that must get approval from the client after the first attempt.

MIX EVALUATION—REJECT A SUBPAR MIX OR PROJECT

This is an important point, as evidenced by previous mentions in the book. A Mastering Engineer must develop enough experience and confidence with recordings and mixes to assess mix quality and viability for mastering. Mastering remains a downstream process, and although impactful changes can be made, the quality of your work is directly affected by the quality of the mixes you receive. If the mix has instrument or frequency balance concerns (among other issues listed ahead), you are wise to request a revised mix and provide the client with specific mix notes. It is helpful to keep an updated list of professional mixers or engineers handy for referral if necessary. By evaluating the mix with *objective* and *subjective assessments* (Chapters 5 and 6), a clear concept of the overall quality of the mix will emerge. Following, I've listed eight common mix issues that may require a revised mix.

1. **Too Loud or Too Quiet:** *Standard mix level* should be between +4dBu and +8dBu on a VU meter, and/or between −6dBFS and 0dBFS (decibels full scale). If the mix is much louder than this, it may be lacking dynamic range or suffer from extreme peak-limiting.[1] If it is too quiet, it may have excessive noise (as the signal-to-noise will be compromised), especially once level is increased in mastering. Also, a quiet mix file may not make complete use of the particular digital word resolution (at 6dB per bit).[2]

2. **Instrument Balances Off:** This occurs if the natural presentation of a group or ensemble is incorrect, or if prominent mix elements or instruments are masked or difficult to discern. Most obviously, if any instrument is interfering with the lead vocal, that must be rectified. If instruments are competing for the exact same frequency ranges, or are panned directly on top of each other, this can also necessitate instrument balance changes and a mix revision.

3. **Frequency Balances Off:** This usually occurs if there are issues with the acoustics of the mix room, or with the speakers. The result is an overly dull or overly bright mix. This issue is generally broadband, affecting either low, mid-, or high frequencies, but could also apply to specific instruments being over-equalized

in the mix. Another example is if the mix engineer boosts similar frequencies on most of the instruments, resulting in an excess of that tonality.

4. **Too Mono or Too Stereo (i.e. Too Much Widening):** A great mix ideally makes effective use of the stereo image. If the mix elements are not properly panned, or conversely, if they have excessive widening effects applied, mastering will be compromised. The obvious exception here is deliberately or historically mono mixes.

5. **Improper Vocal Placement and Presentation:** Obviously, the lead vocal in most recordings is critically important. The treatment of the vocal as broad-band frequency-wise, present, and natural sounding is ideal. The mix should be revisited if the vocal is in any way too quiet or too loud, or masked by competing instruments or frequencies, or if the diction of the words is difficult to follow.

6. **Genre-specific problems:** This refers to inappropriate mix treatments considering the musical genre. Examples might be a rock track with boomy low frequencies and insufficient mid-range; a hip-hop track with thin-sounding kick and bass; a singer-songwriter, classical, or jazz track with excessive limiting. Each genre of music generally dictates specific instrument and frequency balances, along with dynamic range expectations.

7. **Macro-dynamic issues:** I also refer to this as *inverted dynamics*. This is generally an affliction of over peak-limited mixes, whereby the quiet sections become louder than the loud sections. The revised mix should have the dynamics between song sections substantially restored.

8. **Production Shortcomings:** If the quality of recorded sounds or instruments is lacking, or if possibly the expertise of the production team is limited, it will be reflected in the mix. Addressing and revising these concerns with a more established mix engineer can result in substantial improvements.

Make sure to get the highest quality mix available of the recording before beginning the mastering session.

OSCILLATOR SWEEP THE SIGNAL PATH REDUX

I refer to this procedure in Chapter 7—The Mastering Game Plan, but it bears repeating as a valuable tip. Once the mastering chain is selected, it is essential to sweep the chain with an oscillator tone from the PBDAW between at least 20Hz–20kHz and verify that both level and frequency response are flat at the dBFS and VU meter *reference level* of your studio (usually –14dBFS = 0VU = +4dBu) while reading the RDAW output. This is very important, as certain load interactions between analog equipment circuits may have undesirable interactions that can result in level loss, low- or high-frequency roll-offs, or both.

If there are level drops or frequency roll-offs, troubleshoot first by bypassing analog equipment one at a time (notating the result) until you achieve a flat reading. Next, swap the sequence of analog equipment until the response is flat. You may need a technician to add buffer amp stages between the equipment in question, which will solve loading issues. Otherwise, use a mastering console (i.e. the Dangerous Liaison or Maselec MTC-1X) with buffered insert switches for each analog piece to achieve a flat response through the mastering chain.

It is critical that your mastering chain is flat so you are not compromising audio and unwittingly using a mastering system that works against the audio. You never want to add extra level or over-equalize to achieve the desired result; ideally, the quality of your signal path alone adds fidelity to the audio even before any adjustments.

SPECTRUM ANALYZE THE MASTERED AUDIO

Check the *frequency response curves* of mastered audio you like (highly regarded or favorite releases, or your own or a colleague's work) with a *spectrum analyzer* (which implements fast Fourier Transform methodology [FFT])[3] and compare it to an analysis of your current mastered audio (see Figure 5.8 for the Voxengo SPAN spectrum analyzer). This comparison will reveal any frequency response differences that can be addressed during the mastering session. The analysis should also corroborate what you are hearing, and serve as a quick test that your playback system and room are reproducing audio accurately. Unless one is deliberately compensating, a frequency imbalance from the speakers or room acoustics will be inversely represented in the finished audio. For example, bright speakers or a reflective room will result in a dull mix or master, and dull speakers or a dead room will result in a bright master. Your room must be true (flat) to avoid these problems. A *frequency analysis* of program material will help catch this phenomenon if you are adjusting to new speakers or a new room, or have made other changes that affect the sound presentation.

WHEN IN-THE-BOX (ITB) MASTERING IS YOUR FRIEND

ITB mastering is explored in Chapter 16, but I mention it here as a practical alternative to an analog or hybrid mastering chain. If the feedback from your first pass of mastering is less than enthusiastic, or the client otherwise prefers the flat mix, you can do a revision pass ITB. I've found that often the client will enthusiastically approve the revised mastering because ITB mastering offers a gentler *mastering footprint* by avoiding *DA* and *AD conversion* concerns, and extra electronics or compatibility issues that can occur with analog mastering chains. Additionally, by utilizing only a compressor, one or two bands of EQ, and a transparent peak-limiter, the processing is contained, which preserves the characteristics of the flat mix.

WHAT IF YOUR CLIENT IS DIFFICULT?

In rare occasions, it helps to have a plan to manage an unusually difficult client. If they can't let go of a project, or describe phenomena that you don't hear or agree with, explain that you believe the master is fine, but you will indulge a final revision. Avoid endless mastering revisions on vague or contradictory feedback. If you have an office staff or a booking manager, have them alert the client that revisions are limited unless they agree to pay for ongoing work. Fortunately, this situation remains uncommon, and usually, my clients are spot-on with their revision requests. Ideally, they simply love the result and can't believe their ears. However, if the project is not progressing, you can refer a colleague or issue a refund and move on to the next project.

CONCLUSION

These five tips can help a Mastering Engineer maintain a high degree of execution and avoid mistakes. By combining aesthetic musical sensibilities with a repeatable system of approaches and checks and balances, the probability will become very high that your clients will love your mastering work and you will catch issues before the approval stage. Preventing mistakes and managing client concerns with poise becomes critical in building and maintaining a mastering client base. The extra attention is well invested if it avoids the disappointment of an oversight.

NOTES

1. Extreme peak limiting can skew these objective assessment readings, so I am excluding that possibility here to streamline the discussion.
2. Ken C. Pohlmann (2011) *Principles of Digital Audio, 6th Edition*, New York, NY: McGraw-Hill. pp. 28–35: The number of quantization intervals is determined by the number of bits (resolution) available in the digital word. . . . A maximum peak-to-peak signal presents the best case scenario because all the quantization intervals are exercised. . . . As signal level decreases fewer intervals are exercised.
3. Ken C. Pohlmann (2011) pp. 356, 652–653: Jean Baptiste Joseph Fourier first established this relationship between time and frequency. The Fourier transform maps a time-domain function into a frequency-domain function to generate the spectrum of a continuous signal.

Professional Mastering Process — Micro Considerations

Part V begins by revisiting foundational concepts from Chapter 7—The Mastering Game Plan in additional detail, then expands into a vast array of *advanced mastering techniques*. Note that it is beneficial to spend considerable time working with the *fundamental mastering tools* that comprise The Primary Colors of Mastering (EQs, compressors, and BWLs) introduced in Chapter 4 before tackling *advanced mastering* techniques presented in Chapter 14. Part V includes Chapter 15—Mid-Side, an in-depth exploration of this now ubiquitous process, which is a standard option for many modern plug-ins or is included as a feature on modern mastering consoles. The final chapters delve into in-the-box (ITB) mastering, and de-noising and audio restoration. I continue expounding on the *professional mastering process* with a look at *micro considerations*.

CHAPTER 13

Detailed Guidelines for Processing Audio

SELECT CONVERTERS AND SIGNAL PATH

It's important for a professional Mastering Engineer to research, select, and implement *AD* and *DA converters* for their studio. Budget permitting, they may keep an alternate set of converters, especially *AD converters*, for different audio dimensionality, overall sound quality or other features. As presented in Chapter 3—The Anatomy of a Professional Mastering Studio and The Three Zone Mastering System (see Figure 3.9), the basic structure of a two-DAW mastering system includes a PBDAW and an RDAW, each with an I/O sound card or audio interface allowing for Audio Engineering Society (AES)[1] standard digital audio transmission at the output/input of each DAW. The DAWs must be equipped with the software necessary to play back, record, and process digital audio. The *DA converter* after each DAW is connected to a monitor position on the monitor controller. This allows for quickly monitoring the flat (or plug-in processed only) audio from the PBDAW, or switching to the mastered audio from the RDAW *DA converter*. The *DA* and *AD converters* establish the boundaries of The Three Zone Mastering System, which I have conceptualized in order to organize, communicate, and manage gain structure and audio processing through the mastering system.

As discussed in Chapter 7—The Mastering Game Plan, next select the equipment for the analog *signal path*—the analog equipment between the *DA* and *AD converters*. There are numerous options for an analog *signal path*: the type and number of equipment pieces, implementing buffer amplifiers between each piece, mastering console or point-to-point configurations, and the overall sequence are the realm of research and experimentation under the purview of the Mastering Engineer. Achieving an ideal *signal path* that embodies all aspects of high fidelity audio processing is an endeavor that can span years, so keep detailed notes of your best-sounding, most effective mastering setups.

EQUIPMENT SEQUENCING CONSIDERATIONS

Equipment sequencing was introduced in Chapter 7, and is the realm of ongoing research and experimentation for a Mastering Engineer. If you work in a mastering facility or are inclined to research online, experience sharing among colleagues remains very valuable. Seek to participate in a community of other Mastering Engineers if possible. Along with equipment selection, the specific *equipment sequence* is one area where you can fine tune *your sound*—specific sonic characteristics that reveal and accentuate the audio in a pleasing way—creating ongoing demand for your mastering and praise from your clients.

THE THREE ZONE MASTERING SYSTEM (DETAIL MOVES)

In Chapters 3 and 7, I introduce and elaborate on relevant aspects of The Three Zone Mastering System (see Figure 3.9 and Table 7.1). Within these *three zones*, the analog hardware, digital hardware, and/or plug-ins may be sequenced in myriad ways. Following is an outline of the *three zones*.

1. PBDAW (*zone 1*)—before *DA conversion* (digital domain—at source file resolution or can be *sample rate converted* [SRC] to high-resolution)
2. **Between the DA and AD Converters** (*zone 2*)—analog domain
3. **BWL and RDAW Section** (*zone 3*)—post-*AD conversion* and (digital domain—high-resolution)

Characteristically, *zone 2* is the attention-grabbing area of the *mastering system*. Rightfully so, as this is where high-profile, vintage, high-money boutique, custom-designed and/or hand-built equipment lives. However, never underestimate *zone 1* adjustments (optimizing mix level, or preliminary adjustments) for a 'running start' at your analog chain. This allows the analog equipment to do what it does best—not surgery, but tonal sound shaping, and accentuating depth, dimension, detail, and impact (and most of The Eleven Qualities of Superb Audio Fidelity from Chapter 2—Listening Experience). Also note that critical final touches can be executed in *zone 3* as well. Following, I explore *zone 1* moves in more detail.

Zone 1 *Moves—PBDAW Adjustments*

Set Optimal Mix Playback Level

The first task to accomplish in *zone 1* is to achieve an *optimal playback level* through the mastering system. A *standard mix level* is between +4dBu and +8dBu on a VU meter. This means the peaks of your audio should ideally not swing the VU meter needles much beyond +8dBu. 0VU = 1.23V = +4dBu represents the standard calibration level of analog mixing consoles. The dBFS Peak *objective assessment* measurement of the mix indicates how much *zone 1* headroom is available in the PBDAW's software (usually Pro Tools). Never clip the buss in the PBDAW. It is a safe and appropriate first step (after listening and taking measurements) to turn up the faders a few tenths of dB shy of your dBFS Peak reading. For example, if the *objective assessment* measurements are: dBFS Peak = –4dB, dBFS RMS = –20dB, and VU meter Peak = +6dBu, then simple math indicates 4dB of headroom (or +3.9dB) without clipping the buss. I highly recommend using a VU meter plug-in (I particularly like the PSP VU3, as it resembles the look and mimics the ballistics of a real analog VU meter, and is fully adjustable), or an actual analog VU meter.[2] In my experience, flat mixes played back from *zone 1* provide optimal source level through the *mastering system* at a playback VU meter reading between +4dBu and +8dBu. Higher levels risk artifacts occurring through the mastering chain such as distortion, pixilated high frequencies, or overloading at the inputs of downstream analog devices. For most mastering applications (and for a number of years of regular mastering), only perform *level adjustments* in *zone 1* of the mastering system. Introducing other adjustments is acceptable once the fundamentals are mastered. Another important detail to point out for *zone 1* is that many DAWs perform *digital signal processing* (DSP) functions at 32bit or 64bit processing precision, meaning the moment level is adjusted the system reverts to the internal DSP precision of the DAW and then should be *dithered* back to 24bit (the highest bit depth conversion of *DA converters*) to avoid any truncation of the digital word. There are many *dither* plug-ins available, but I regularly use the PSP X-Dither (Figure 13.1), which is excellent.

Write Macro-Dynamic Automation

The second signal processing option available in *zone 1* (the PBDAW) is *macro-dynamic automation* (Figure 13.2). This refers to volume changes between song sections (intro, verse, pre-chorus, chorus, solo, etc.) to preserve the natural cadence of the

Figure 13.1 The PSP X-Dither is a mastering dither and noise shaping processor.

Source: (Courtesy PSP)

Figure 13.2 Example of macro-dynamic automation—slight volume changes by song section on a flat mix file in the PBDAW. The dip before the chorus helps accent the leading edge transient of the chorus downbeat.

recording. *Macro-dynamic automation* helps offset excessive dynamics loss from downstream compressors or limiters that may offer level and a sound you love. Ultimately, it allows for the preservation or even accentuation of the payoff of a chorus or hook described as *blossom* in Chapter 2—Listening Experience. Always avoid delivering an over peak-limited 'brick' of audio. Remember, terrestrial radio stations use broadcast compressors that further peak-limit mastered audio, and most online streaming services such as Spotify or Apple Music have LUFS loudness specifications for quieter master files, so the audio is optimized for their platform.

Preserve Leading Edge Transients

While adding automation adjustments, preserve *leading edge transients* at the downbeats of choruses or other sections. A quick volume ride down, then up to the chorus level, supports the kick drum and other instruments at the downbeat of the chorus. This move has the most impact when going from a slightly quieter section to a louder (payoff) section. In many instances, you may not need to write automation adjustments at all. The discretion that experience provides the Mastering Engineer informs their judgment regarding when to do less (or nothing), when to adjust the track more, and how not to harm the recording in either instance.

Advanced PBDAW Adjustments

Only perform the adjustments described ahead if you are confident that the audio is not being compromised due to over-processing. As mentioned earlier, spend ample time mastering audio with only The Primary Colors of Mastering (a compressor, EQ, and a peak-limiter) so that you have experience making mixes sound great with only three devices. Practice this to improve your fundamental mastering skills and become expert at those three devices. Begin experimenting with these *advanced mastering techniques* on test audio before you are in a session, especially a client-attended session. It's ideal to avoid encouraging suspicion while you are mastering! Of course, your skill level should be such that your clients trust your audio judgment, but know that you may need to explain the procedure if a client becomes curious. If you feel confident with the fundamentals of mastering, exploring advanced *zone 1* adjustments in the PBDAW is a natural evolution. One reason for caution is the understanding that with every additional procedure, the margin of error—the possibility of something being setup incorrectly—increases. Time spent mastering engenders the 'triple-check care and patience' required to execute more complex mastering setups without realizing hours into a project that the session was setup incorrectly . . . read: start over! If the goal is to become an accomplished expert, there is no substitute for time spent regularly mastering projects. Following is a list of additional *zone 1* processing to experiment with:

> Compression—If appropriate, preliminary compression can be added for a running start on a 'radio' or 'hyped' sound.
> EQ—Can be implemented to enhance, fill in or remove tonal aspects of the mix that are not ideally balanced. A professional Mastering Engineer expects to receive an excellent flat mix. This is not always the case, so some adjustments can be made within reason. If the mix is really troublesome, as always, send it back with mix notes.

Parallel Processing—Keep an ear out for any phasing anomalies while parallel processing, as by definition, it means that a duplicated and alternately processed stereo pair of audio plays along simultaneously with the original source audio. Remember to turn on *delay compensation* in Pro Tools if doing this in the PBDAW. Primarily compression and/or EQ are suitable for parallel processing in mastering.

De-Essing—The client may expect a fairly loud payoff volume in the loud song sections. Unfortunately, the byproduct of this is more apparent high-frequency information. Enter the de-esser. Loud mastering work can benefit from a transparent de-esser (good options are the Brainworx Dynamic EQ and UAD Precision De-Esser). This allows the Mastering Engineer to preserve some listening comfort in cymbals and vocal sibilance at the final *target level*.

Zone 2 Moves—Analog Signal Path

As mentioned previously, *zone 2* is where the coveted mastering equipment lives, and can be considered the heart of the mastering system. Many fidelity enhancing adjustments can occur here: tonal shaping with EQs, increasing apparent volume with a compressor, adding detail with parallel processing, or introducing warmth or any desirable quality imparted by discrete analog components. Tubes, transistors, resistors, capacitors, and op-amps in analog equipment circuits can add desirable qualities to the audio such as detail, frequency enhancements, or warmth. *Zone 2* analog equipment not only performs its designated function, but also can impart these other appealing qualities. For example, an *electro-optical* or *variable-gain* type of compressor design may add a pleasing extension in the low or high frequencies—meaning you can use that device as a de facto EQ, in addition to it being a compressor/limiter. You may find that the gain on some analog devices may add mid-range, and also some saturation to the master; or other devices may remain flat when used to add gain—evenly boosting all frequencies. These observations about equipment are critical to mastering decisions, and help determine what to leave in the chain, and what to leave out.

The Mastering Console (Gain and Other Features)

If a *mastering console* is implemented into *the mastering system*, it will be in *zone 2*. The input and output gain controls of the *mastering console* provide strategic gain positions that can be used to adjust gain structure through the mastering system, drive a compressor harder, balance the left/right (L/R) image, or drive (clip) the *AD converter* (or a pre-converter clip circuit). Driving or clipping the *AD converter* is covered in more detail in Chapter 14—Advanced Mastering Chain Tools. It remains a common method used by Mastering Engineers to achieve a louder master, but note that extreme care is in order; familiarity with the *AD converter* is critical, including exactly how it overloads and distorts. I use the Dangerous Master/Liaison, which is a *mastering console* setup providing great flexibility in *zone 2* of my mastering system (see Figure 3.7). It also offers a *Mid-Side* feature for strategic widening. Other common analog devices beyond EQs and compressors suitable for *zone 2* are de-essers, parallel enhancements (many mastering consoles are equipped with a parallel fader), and clipping/saturation devices. This means that as an *advanced mastering technique*, even if parallel processing was done digitally in *zone 1* (PBDAW), it may also be utilized in *zone 2* after the *DA converter*.

Zone 3 *Moves—After AD Conversion and in the RDAW*

As discussed in Chapter 4—Fundamental Mastering Tools, the classic mastering device placed post-*AD conversion* is the *brickwall limiter* (BWL). This is a digital look-ahead peak-limiter that allows you to set a dBFS threshold over which no audio signal will pass. I commonly set my BWL ceiling for –0.8dB. This maintains a breath of headroom and accommodates the additional level that *dither* will add after rendering to 16bit—44.1Hz, and also allows for a modicum of headroom for creating an MFiT compliant file if requested.

Other *zone 3* adjustments include additional *macro-dynamic* editing at the RDAW to preserve intro or verse-to-chorus dynamics; de-essing, which could either be a high-quality DSP device (Weiss digital hardware or t.c. electronic M6000) or you can select a plug-in de-esser on spattery vocal passages, edgy cymbals and sections with excessive high-frequency information (try 'spot' de-essing if most of the track sounds fine); and some problematic source mixes may still need some surgical touch-up EQ to get the track to sit perfectly. Note that when processing or adjusting a file that has already been recorded into the RDAW, this will result in over-levels dBFS, meaning that you must follow your *zone 3* processing plug-in(s) with a transparent BWL plug-in, set in the range of –0.8––0.4dBFS.

Adjustment Sequence: Gain Structure, Dynamics, and EQ

In Chapter 7 under the heading Execution of The Mastering Game Plan, I present a prioritized sequence of mastering adjustments (work priorities) to execute for each song once the *objective* and *subjective* assessments are complete. These are:

1. Set the *gain structure* through the mastering system for the chosen *target level*
2. Make *dynamics* adjustments next to manage peaks or add *apparent volume*
3. Make *EQ* adjustments last and begin with low frequencies first

Performing these adjustments for each song in an album or collection helps create a cohesive result, and keeps the mastering session progressing. It provides a substructure for the session and gives the Mastering Engineer additional confidence and command over the process. Please note that subtle adjustments can occur in any order, but this sequence allows the engineer to establish the *target level*, *sonic impact*, and initial *EQ* settings systematically.

The *advanced mastering techniques* that will be introduced and explored next in Chapter 14 also fit into this sequence of adjustments depending on the technique. If the technique adds gain, it is done first as part of *gain structure*; if it affects *dynamics* (i.e. parallel compression), it is done second with other *dynamics* adjustments; and if it affects *EQ* or frequency response, it is done third, along with other *EQ* adjustments.

Match Mix and Master VU Metering

Once all mastering adjustments are complete, and you are recording the mastered file into the RDAW, verify that mix and master VU metering match. The plug-in VU meters (PSP VU-3) in *zone 1* (on the PBDAW) should match—have similar ballistics to those of—the analog VU meters reading *zone 3* (after the RDAW). The difference is that the 0VU *reference level* at the *zone 3* (RDAW) VU meter will be 4dB–6dB higher to accommodate the louder level and peaks of the mastered audio. This verifies a fundamental goal of successful mastering: to preserve original transient response and *macro-dynamics* despite added compression, gain, or EQ. Confirm that intros and verses of the master remain generally quieter and less impactful than choruses or payoff sections. The first sign of subpar mastering is a master that is carelessly slammed through a peak-limiter with little attention to *frequency balances* or *macro-dynamics*.

CONCLUSION

This chapter represents a detailed look at the mastering process and the sequence of adjustments that are made as audio is processed and optimized. It bridges concepts discussed in Chapters 3, 4, and 7 for a relational look at The Three Zone Mastering System, Fundamental Mastering Tools, and The Mastering Game Plan.

NOTES

1. Ken C. Pohlmann (2011) *Principles of Digital Audio, 6th Edition*, New York, NY: McGraw-Hill. pp. 488–489. The AES3 standard interconnection permits transmission of two-channel digital audio information.
2. A digital *loudness units full scale* (LUFS) and/or *loudness units* (LU) meter can also be used to take *objective assessment* measurements. There are numerous options for LUFS/LU measurements, including Steinberg WaveLab DAW software, FabFilter Pro-L, and DMG Limitless plug-ins.

CHAPTER 14

Advanced Mastering Chain Tools and Techniques

This chapter explores advanced mastering methods. These techniques are more complex and nuanced, involve a greater margin of error, and require the confidence of execution that only experience provides. Used skillfully, a Mastering Engineer can achieve impressive, high-quality results. As always, it is critical to lean on the fundamental habit of referencing your flat mix religiously so that your adjustments represent a genre-appropriate optimization rather than a dramatic reinterpretation of the mix.

SIDE-CHAIN COMPRESSION (S.C.)

Most mastering compressors are equipped with an onboard s.c. function or external s.c. input (or both). This is so a 'mult' (duplicate) of the program material can be EQ'd and sent to the s.c. input, thus affecting the frequencies the compressor does or does not react to. Two popular examples are a high-pass filter, so that the kick drum does not cause excessive compression or pumping (the audible attack and release of the compressor)—often onboard in mastering compressors; and a de-esser, whereby a high-frequency boosted version of the song is sent to the s.c. so that the compressor reduces those same frequencies. Another s.c. option is sending a pre-compressed version of the audio to the s.c. input so that the compressor functions more gently on transients, even with fast attack times. Examples of quality mastering compressors with side-chain functionality are the Manley Variable-Mu™, Manley SLAM!™, Pendulum Audio OCL-2, Shadow Hills Mastering Compressor, Magic Death Eye Mastering Compressor, and Alan Smart C2.

MULTIBAND COMPRESSION/LIMITING (MBC/L)

This is a compressor or limiter that separates the audio spectrum into separate frequency bands that can be adjusted independently of each other. Imagine each band as its own independent compressor or limiter with its own threshold, ratio, attack, and release. Some companies even include harmonic saturation/distortion and also

M/S functionality. A multiband dynamics processor can change the instrument and frequency balances significantly, so I'm careful not to overuse them and alter the mix excessively, imparting too drastic of an imprint on the source audio. A great mastering job is always relatable to the flat mix and should be an enhancement of what the recording and mix engineers sought to achieve. I would recommend MBC/L for mixes that are problematic, or for genres that require an extreme 'hyped' result—sometimes found in pop, dance, electronic dance music (EDM), electronica, or possibly some sub-genres of hip-hop/rap (Figures 14.1–14.2).

Figure 14.1 The UAD Precision Multiband offers five spectral bands of dynamic range control. Separately choose compression, expansion, or gating for each of the five bands to manage everything from complex dynamics control to basic de-essing.

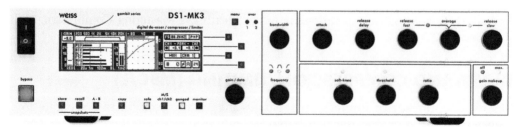

Figure 14.2 The Weiss DS1-MK3 is a standalone digital hardware dynamics processor that handles de-essing (with low-pass, band-pass, and high-pass selectable compression band), compression, and limiting functions.

SERIAL PEAK-LIMITING

This technique involves the risky outcome of a loud result with very diminished dynamic range and works most effectively on sparse but punchy music such as rap/hip-hop, dance/electronica, or EDM. Serial peak-limiting involves the extremely bold move of setting up two peak-limiters in series either as plug-in instances, engines in the t.c. M6000 hardware, actual L2 hardware units, or a combination of these options. The peak-limiters are in *zone 3*, after the AD converter and each one generally adds approximately 1dB of gain below threshold. The output ceiling can be set to taste, but I generally use two different plug-ins with a ceiling of –0.3dB for the first peak limiter and a ceiling of –0.6dB for the second one. Listen very carefully to the effect of each limiter, and if the audio fidelity deteriorates, remove one or both.

DYNAMIC EQ

Although I primarily use dynamic EQ in mastering for *zone 1* de-essing, it is worth exploring. Like MBC, for mastering applications, dynamic EQ may help with some surgical fixes on specific elements or for additional tonal balancing within mixes. A dynamic EQ combines the band-separated aspect of a multiband compressor with the linear changes provided by an EQ. As such, in addition to setting EQ filter parameters (frequency, boost/cut, shape, and Q), there are dynamics parameters such as attack, release, threshold, and even s.c. options. This means the selected EQ is interacting with the dynamics of the music and engaging as needed. There are several good plug-in examples available: brainworx dynEQ_V2, The Ozone 8 Dynamic EQ, and Hofa IQ EQ3.

SATURATION

Saturation is a type of gentle distortion that can add some vitality and dimensionality to a mix that is lifeless or contained-sounding. Saturation is a byproduct of overloading the input or output of analog equipment like tape machines, tube amplifiers, or console microphone preamplifiers to create harmonic distortion. The increased input amplitude exceeds the capacity of the electronics or analog circuit to cleanly pass the signal. This can result in a richer and warmer sound, up to a point . . . beyond which the mix will sound overly distorted.

There are digital emulations of saturation as a feature in both hardware devices and plug-in software. For example, the t.c. electronic Finalizer includes a DRG (digital radiance generator) function allowing the user to dial-in emulated saturation, and the Crane Song HEDD-Quantum AD converter has options for tube and tape saturation emulation. In addition, the DMG Equilibrium plug-in has a 'harmonics' option for either a high-pass or low-pass filter that saturates the unfiltered (remaining) frequencies in the audio spectrum, and following I will review tape saturation plug-ins, as well. These are effective tools to have in your mastering arsenal, but be very careful with saturation as you are deviating further from the qualities of the flat mix.

ANALOG TAPE COMPRESSION

One original reference point for saturation is *tape compression*, whereby the input signal exceeds the capacity of the metal oxide particles in the tape to cleanly reproduce the given amplitude. This results in the tape becoming saturated, sounding like a combination of both compression and subtle saturation. The standard analog playback and record tape machine alignment settings—repro level, record level, record bias,[1] high-frequency and low-frequency adjustments—allowed recording engineers to align the machine to match tape manufacturer specifications, or to elevate the record gain adjustment and 'hit the tape harder' for a louder, punchier, and more saturated sound. With the advent of digital audio, this time-consuming and expensive analog process is available virtually in the form of various tape machine emulation plug-ins. Popular options are Slate Digital VTM, Avid Reel Tape, and UAD Magnetic Tape Bundle (which includes both Ampex and Studer virtual tape machine options). Of course, if you want to be a 'mastering cool kid,' you can get an actual analog tape machine.

If you have access to a good analog two-track machine, some quality recording tape and a bit of time, you can transfer a digital source file to analog tape before mastering to add warmth and dimensionality via tape compression. Keep in mind that analog two-track was the mix engineer's format of choice for several decades. In my comparisons, the Ampex ATR 102 two-track machine sounds 'quicker' reproducing transients, making it great for rock with live drums and pop styles of music. The Studer A820 two-track machine—which handles tape superbly—seems to round out certain transients and sounds 'slower' on the reproduction of drum transients.[2] The 15 inches per second (ips) tape speed shifts the machine's 'head bump' (low-frequency response buildup on playback from the tape machine) lower and creates additional richness or warmth. Tape at 30ips is generally a bit more 'glassy' and extended sounding in the high frequencies, but still a great option for an extra analog element in the mastering process.[3]

My favorite configuration is Ampex 456 tape at a +6 record level alignment[4] at 15ips on an Ampex ATR 102 two-track tape machine with a half-inch head stack. I then print three separate samples of the song—usually about one minute of a chorus—at three different levels (increasing in increments of 2dB) to play back and make a final determination of which level sounds best and presents the most appealing tape compression sound. Remember to not get seduced by the process and to always use your best judgment. This means that sometimes the digital file is still the clear winner.

CLIPPING

Another method to create a louder or more aggressive result is called *clipping*. The simplest way to do this is to increase the last gain stage before the *AD converter* thereby *clipping* the converter for a louder result accompanied by characteristics slightly harder than saturation. Telltale signs of *clipping* are squared-off peaks in a waveform. Another method is to have an analog clipper circuit in your signal path placed just before the *AD converter*. This allows you to push gain and achieve apparent volume, as well as protect the *AD converter* from producing unpleasant artifacts from overloading. These use a discrete component circuit of transistors, Zener diodes, or even LEDs in the feedback loop of an op-amp. The clipping can be designed to be either a softer or more rounded response at the +18dB point for a gradual or warmer sound, or harder and more angular, yielding a harsher more quickly distorting sound. Common clippers are the Prism Sound 'Over-Killer' barrels (a diode circuit that clips signal voltages above a +18dBu level) and the Clip +18 setting on the Manley SLAM! The +18 clipper is designed to function in a mastering system that is referenced to −14dBFS at 0 VU = 1.23 V = +4dBu (if −14 = +4dBu, then +18 will be at 0dBFS at the AD converter) (Figure 14.3).[5]

Figure 14.3 The Magic Death Eye Evrenizer is a one-off (serial #0001 of 1) analog clipper that gradually switches between soft and hard clip settings.

Analog clippers will clip and eventually distort, and even pass signal above the clipping threshold. In contrast, digital BWLs will not pass signal above 0dBFS. The original Apogee AD series converters had a soft limit function to prevent over-levels near 0dBFS. At Capitol Studios in the mid- to late 1990s, the tech staff under Jeff Minnich[6] built the fabled 'black box,' which the technical staff originally made for Mastering Engineer Wally Traugott to create hotter masters.[7] It was a clipper that would remove signal above a certain threshold. As it was an analog device, if too much level was sent through the unit, it would pass over-levels, so the digital BWLs (such as the Waves L2 or the t.c. electronic M6000) provided a cleaner and more reliable option.

PARALLEL PROCESSING (COMPRESSION AND EQ)

Parallel processing refers to audio signal that is multed then processed—usually more aggressively than normal—and then blended in with the original stereo signal. You must always verify there is no delay or latency between your main program and the parallel program while doing this, or the result will be much worse than the flat mix.

As I consider parallel EQ as generally corrective, I perform it in *zone 1* of my mastering system, meaning the PBDAW. This is also due to the incredible set of surgically adjustable parameters in many mastering EQ plug-ins. My approach for tonal balancing is to set up parallel EQ with a certain band of frequencies boosted and the remaining bands cut quite aggressively or high/low-pass filtered. If a mix is light in the low frequencies, for instance, I can blend in a low-boost stereo pair to compensate. Or, if the vocal is lifeless, I can blend in mid or high-mid frequencies until they just begin to affect the detail of the vocal. What you adjust affects everything in mastering, so consider M/S EQ for your parallel work to minimize the overall footprint (Figure 14.4).

For parallel compression, I prefer the sound of analog compressors and use the analog blend fader on my mastering console to add in detail from a compressor set for 4–8dB of gain reduction, which would be considered maniacal in a standard mastering context. Be careful with transformer or tube-designed compressors for parallel applications, as they may introduce a very slight delay that will smear the audio image. Many dedicated mastering consoles allow for parallel functionality with a stereo blend knob/fader such as the Dangerous Master and Liaison, Maselec MTC-1X, SPL DMC, and Crookwood Mastering Consoles.

MID-SIDE PROCESSING (M/S)

M/S Processing represents an important advanced mastering technique which I cover in detail in Chapter 15—Mid-Side—An Elixir of Mastering Hope. The methodology originates from the Blumlein stereo recording technique, and allows for mix elements in the middle of the stereo image to be processed separately from elements in the sides of the image.

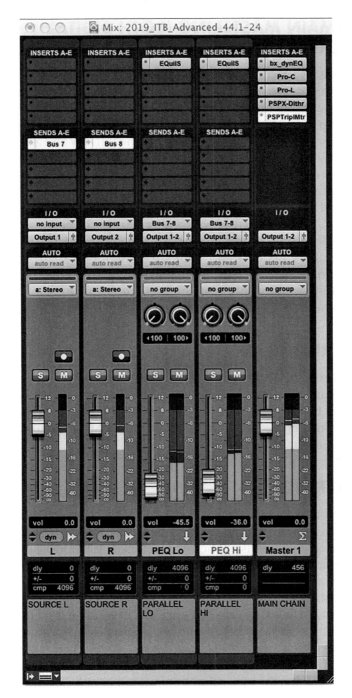

Figure 14.4 An ITB ProTools mastering session setup up with low and high-frequency parallel EQ blended in before being routed to a mastering chain on the master fader. This could also be an advanced PBDAW *zone 1* setup before converting to analog for *zone 2* and then capture in *zone 3*.

HYBRID DOMAIN MASTERING (BOTH DIGITAL AND ANALOG)

This represents less of a technique and more of an understanding of increased complexity in the *mastering system*. As opposed to either performing ITB mastering, an all-digital hardware chain, or a primarily analog chain, a *hybrid domain* system incorporates all of them. As outlined in Chapter 3—The Mastering Studio, I separate the *mastering system* into three distinct zones: *zone 1* is the PBDAW and any digital processing before *DA conversion*, *zone 2* is the analog equipment chain in the *mastering system*, and *zone 3* is the region after *AD conversion* where any digital hardware processing (DSP or peak-limiting) and then any RDAW adjustments occur. Incorporating aspects of both digital and analog processing represents the best-sounding fidelity for most mastering projects. There are certainly exceptions, as I will discuss in Chapter 16—In-the-Box Considerations. In my many tests and comparisons, The Eleven Qualities of Superb Audio Fidelity from Chapter 2—Listening Experience remain most enhanced in a *hybrid domain* system. Note this does not require excessive processing in each *zone*—at times simply setting up the *gain structure* to achieve the desired *target levels* is all that is necessary.

LINE AMPS AND TRANSFORMERS

Mastering Engineers occasionally seek to add *line amps* or *transformers* to their analog *signal path* for alternate gain staging, coloration, vitality, or light saturation. These devices are custom built by the Mastering Engineer or a trusted technician. Suffice it to say that all Mastering Engineers are on an enduring quest to set their work apart vis-à-vis custom equipment or modifications.

STEM MASTERING

Stem mastering involves the mix engineer separating mix elements into separate stereo pairs. The simplest example is a *vocal stem* and an *instruments stem* that combine to make the final mix. *Stem* configurations can become increasingly complex; for example, using four *stems* (*drum stem*, *bass stem*, *music stem*, and a *vocal stem*) is also common. This approach between mixing and mastering allows for stunning results without having to compromise enhancements (the Mastering Engineer's eternal dilemma). To illustrate, by boosting 4.8kHz, many mix elements are affected—a snare may be enhanced enough, but a guitar may be simultaneously compromised by sounding edgy. With *stems*, for example, the bass frequencies can be enhanced without affecting the vocal, the vocal can be brightened without affecting the cymbals, and the kick drum can be enhanced without making the bass guitar boomy.

    Stem mastering also allows for analog summing of the final mix, as the stems are played out of the PBDAW, through the analog mastering chain, and recorded into the RDAW. However, this approach can increase margin of error if the mix engineer has balanced the full mix through a limiter, then generates *stems* through the same limiter setting. The lower level of each *stem* will not affect that limiter as intended, and instrument balances will be off. It is best to mix without excessive limiting if you will be generating stems for a Mastering Engineer. Although this is not a hard-and-fast rule, most *stem mastering* adjustments occur in *zone 1* (the PBDAW), before routing to the stereo analog chain.

INTERCUTS (MASTERING MOVES BY SONG SECTION)

There are instances when a song section requires a different treatment than the rest of the song. Some examples are an instrument solo that is slightly buried in the mix and needs more high-mid EQ, a bridge that sounds better with more low frequencies accentuated, or a loud climactic ending where cymbals become overbearing and require EQ or de-essing. Although these changes can be attempted 'on the fly' in real time, it is best to go back and record the sections with different settings into the RDAW as an *intercut* and edit them into the original mastered pass of the song.

These surgical mastering moves can make a significant impact in improving the mix and final master. It is imperative to be extra careful with this type of editing and zoom in to verify that each seam of the edit(s) line up perfectly. If there are several *intercuts* in your project, consider rendering the file to permanently capture the edits and name in a new session (while saving the previous edit stage just in case) so they don't inadvertently get moved or altered, causing alarming drop-outs or skips in the master. These edits and captures must be performed on the high-resolution file first captured in the RDAW, maintaining the original *sampling frequency* and *bit depth* (as mentioned, I regularly use 96kHz—32bit float), not the rendered Red Book .wav file (44.1kHz—16bit).

ADDING REVERBERATION

Reverberation or reverb refers to the natural decay of a sound source in a room or hall as the sound waves bounce off of the walls, ceiling, and floor. Early reflections are the initial reflected sound waves. Reverb time is determined by the room volume and absorption. In the early days of analog tape, recording a musical performance was captured live to a mono or stereo tape machine, and if the recording studio was acoustically absorbent (dead), reverb could be added via a live echo chamber to achieve a more natural, acoustically blended sound. The final application of reverb was often left to the mastering session. Later, with the advent of multi-track recording and other reverb options (spring reverbs, reverb plates, digital reverb units), the addition of reverb remained the domain of the mix engineer.

At Capitol, there are eight live echo chambers[8] underneath the parking lot. These chambers were reputedly designed by Les Paul, and constructed initially as a group of four when the building was built in 1956, with four more added soon thereafter (Figures 14.5 and 14.6). Marquee artists ranging from Frank Sinatra and Nat 'King' Cole to The Beach Boys, among many others, have utilized them on legendary recordings. I am able to access these chambers from my mastering studio, and on certain recordings—primarily singer-songwriter, solo instrument, or some sub-genres of rock—it can add a unique vitality and ambience. A *common airspace* on certain recordings imparts a natural and familiar *listening experience*. At any live musical performance, there is a single concert hall or venue that imparts its particular reverberation characteristics to every instrument and

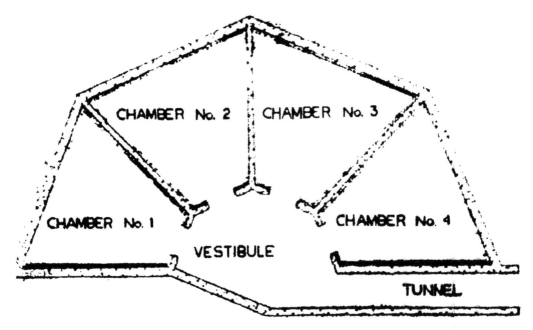

Figure 14.5 The original floor plan of the first four Capitol echo chambers built in 1956.

Source: (courtesy Journal of the Audio Engineering Society)

Figure 14.6 Construction of the Capitol echo chambers.

Source: (courtesy Universal Audio/UAD)

sound. Of course, there are myriad plug-in options for reverberation, as well, and UAD has recently released a plug-in version of the Capitol echo chambers. From a mixer's perspective, adding reverb to a mix at the mastering session may seem problematic at best, but it does have a place in the Mastering Engineer's arsenal of approaches.

CONCLUSION

These *advanced mastering chain tools* are best incorporated into the *mastering system* gradually, if The Primary Colors of Mastering are not providing the desired sonic result or impact. The Mastering Engineer may develop their approach so that two or more of these methods are implemented into the same *mastering system*. It is best to add one technique at a time in order to discover its strengths or weaknesses, and determine how it affects the fidelity of the mastering project.

EXERCISES

1. Set up a mastering compressor with a side-chain input to pass audio with about 2dB of gain reduction. Try three approaches altering the side-chain input (de-essing, high-pass filter, compressed audio). Keep the compressor settings the same except the threshold to isolate the side-chain effect. Notate how the compressor function changes in each instance.

2. Using only an MBC/L or a dynamic EQ plug-in, master a song. Compare the result with the flat mix. Notate the result(s), and explain if you consider it an improvement or not.

3. Add a tape emulation plug-in to the same song and mastering chain used in Exercise 2. Compare the mastered files. Describe how the tape emulation plug-in changed the new master.

4. Set up a simple analog mastering chain (source file–DA converter–EQ compressor–AD converter), then clip the AD converter by increasing the compressor output gain, and record the result. Compare to a recording of the same song and setup *without* clipping the AD converter. Notate differences between the two files.

5. Master a song in ProTools (ITB) with only a parallel EQ setup and a peak-limiter. Describe how you arrived at the final settings.

6. Master a song with a vocal using only a M/S EQ plug-in, a compressor and a peak-limiter. Enhance the vocal only then capture, enhance the kick drum only then capture, finally enhance the guitars/keyboards only, then capture. Play each of the three versions, then describe how you utilized M/S EQ to affect only specific areas of the mix.

NOTES

1. An inaudible high frequency added to the record head to improve frequency response and alleviate low-level distortion. Correlated to record head gap, tape speed, and tape formulation.
2. I have used both of these analog machines throughout my career on numerous remastering projects from original tape sources from Capitol Records and its parent companies' (EMI and Universal Music) archives. Considerations such as tape speed, recording era, noise reduction (usually Dolby type 'A' or 'SR'), and original

or protection copy all affect the audio fidelity of the source tape. Whereas both are effective—the Studer A820 machine excels at handling tape and the Ampex machine sounds punchier with a quick transient response, especially good for rock and pop styles. The condition of the tape heads also contributes to the sound quality, hence head re-lapping and retrofits such as the Flux Magnetics head stacks for the ATR. Custom tube playback circuits are also prevalent in mastering studios.

3. Correlation between tape speed and frequency response.
4. Record level alignment reference.
5. Metering reference levels as a function of voltage and decibels.
6. Jeff Minnich was a technician/chief technician at Capitol Records from 1986–2006.
7. Wally Traugott was a Mastering Engineer at Capitol Records from 1967–1997. Among the many prominent projects he mastered over three decades was the original US vinyl release of Pink Floyd's "Dark Side of the Moon".
8. James W. Bayless (April 1, 1957) Innovations in Studio Design and Construction in the Capitol Tower Recording Studios. *Journal of the Audio Engineering Society* 5 No.2, 71–76. This white paper discusses the Capitol echo chambers, including their dimensions, construction, and design.

CHAPTER 15

Mid-Side (M/S)

An Elixir of Mastering Hope

Mid-Side (M/S) processing represents an incredibly powerful method of enhancing stereo audio by encoding the left and right channels of a stereo recording into middle and side information from the stereo image. This is done via a matrix that encodes the stereo signal into M/S, where it is processed, and then decoded back to stereo. Considering typical panning relationships within the stereo image (see Tables 1.2–1.4 from Chapter 1), M/S processing allows you to isolate and process all mix elements in the center of the image separately from the mix elements that are panned or at the sides of the image. This can be handy for adding air to a lead vocal without making the stereo cymbals edgy; or adding some mid-range bite to electric guitars without affecting the snare and/or lead vocals. Additionally, by adjusting the level relationship between the middle channel and the sides channel, the stereo width of the mix may be adjusted. A multitude of options exist, and well-executed M/S processing can contribute to a very impressive mastering result.

M/S AT CAPITOL MASTERING

When I began mastering in 1995, M/S processing was barely a topic of discussion, and for the most part, the generation of Mastering Engineers at Capitol preceding mine did not implement it at all, except for Bob Norberg.[1] Bob often experimented with mastering approaches and various techniques to process, enhance, and restore audio. He was involved with restoration work of noisy analog master tapes using Sonic Solutions NoNoise and later the AudioCube (an audio restoration DAW). Additionally, his background in recording classical music gave him familiarity with stereo recording techniques, including the Blumlein array.[2] He utilized a method of centering off-center lead vocals while remastering Capitol catalog albums (an affliction of dual-function early stereo master tapes that were mono-compatible; standard centered vocal placement in stereo would pop the vocal level too far forward when collapsed to mono). This was a parallel process that used phase relationships to slightly center the sides. He also implemented a method of widening stereo images, which was the encoding portion of the M/S matrix that he would set up in a Sonic Solutions DAW. He would discuss and document these processes, and I would experiment and use them in my own remastering work for Capitol catalog projects.[3] Then around the year 2000, it became common for Mastering Engineers to find schematics and build their own M/S analog encoder/decoder boxes. Our chief technician at the time, Tom Schlum,[4] built one that I used regularly on mastering projects.

M/S STEREO RECORDING TECHNIQUE—HISTORY AND ORIGINS

The technique dates to the early 1930s and is credited to EMI sound engineer Alan Blumlein, who has the original patent.[5,6] This stereo microphone recording technique uses two coincident microphones: a figure-eight (bidirectional) microphone facing sideways and a cardioid or omnidirectional microphone at a 90° angle to it, facing the sound source.

The left and right channels are produced through a simple matrix on the recording console that creates the stereo image through a combination of signal polarity relationships and signal summing described following.

Microphone Signals

There is one mono signal from the cardioid microphone and one from the bidirectional microphone. With a bidirectional pattern, the two sides are 180° out of phase, so a positive charge to one side of the microphone diaphragm creates an equal negative charge to the other side.

At Console Input

There are three channels used: one for the cardioid microphone input, and the bidirectional microphone signal multed (duplicated) into the other two channel inputs, with one of those channels *polarity-reversed*.

Signal Summing Matrix at Console

In Figure 15.1, the front of the microphone (+ side), is pointed to the left of the sound stage, while the rear (– side) is pointed to the right. The left channel is created by the *middle* (cardioid microphone signal) panned center and combined with the *in-phase side* (front of figure-eight microphone signal) panned left. The right channel is created by the *middle* (cardioid microphone signal) panned center, and combined with the *polarity-reversed side* (rear of figure-eight microphone signal) panned right.

This configuration produces a completely mono-compatible stereo signal, and if the *middle* and *side* signals are recorded (rather than merely matrixed to left and right at the console), the stereo width can be manipulated after the recording has taken place. By substituting stereo program material for the microphone signals, M/S processing can be performed in other applications such as mastering.

IN A MASTERING CONTEXT

When mastering a stereo recording in M/S, the existing stereo signal is *encoded* into *middle* (M) and *sides* (S) components so that that they can be processed or balanced independently, and then *decoded* back to stereo.

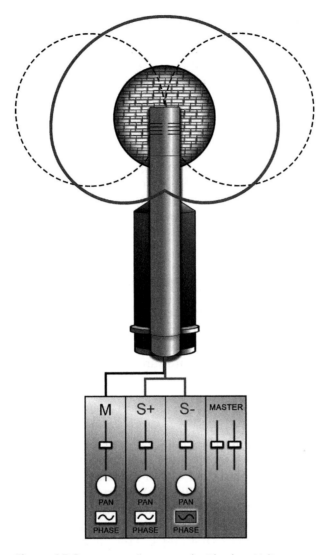

Figure 15.1 Diagram illustrating the Blumlein M/S microphone recording technique.

ENCODING FROM STEREO INTO M/S

In order to create the *middle* channel, combine the left and right channels in phase so the *middle* channel is a combination of left and right in mono (L + R). For the stereo component, combine the left channel with a right channel that is *polarity-reversed* so that the *sides* channel is a combination of left and polarity-reversed right in mono (L − R).[7] This removes the common information between stereo left and right that constitutes the center (*middle*) of the image through *phase cancellation*.

With the audio *encoded* into M/S, a wide range of image-specific processing can occur. This is often isolated EQ on image-specific mix elements or careful balancing of instruments in the image, such as lifting a vocal or enhancing width or definition on stereo guitars or keyboards.

DECODING OUT OF M/S BACK INTO STEREO

Decoding the *middle* and *sides* signal back into stereo is the same process used by commercial M/S matrix systems and can also be accomplished by using three faders of a console: 1) bus the *middle* signal to fader 1 and pan center; 2) bus the *sides* signal to fader 2 and pan left; and 3) bus the *sides* signal to fader 3 with the polarity-reversed and pan right.

The result of this is that the left channel is a combination of M + S; or, substituting formulas from the previous *encoding* descriptions, (L + R) + (L– R) = 2L. Note that the original right channel disappears by combining it with a polarity-reversed duplicate. The right channel is a combination of M–S, or again substituting formulas from the previous *encoding* descriptions: (L + R) – (L – R) = 2R. Notice that the output level is doubled upon *decoding*, requiring you to decrease both the left and right channels by half (6dB) so that the level matches the original audio.

CREATING AN M/S MATRIX IN PRO TOOLS

Figures 15.2–15.5 outline a step-by-step method to set up a M/S matrix in a DAW such as Pro Tools. This is to help gain understanding of the process and grasp how it functions. You would likely not use this setup in a practical application, especially with the plethora of available M/S plug-ins and hardware available today.

ENCODING (DAW EXAMPLE)

Set the stereo source track output to bus 1 and 2. Create two stereo aux tracks, and label them M and S. In each aux track set the input to *both* bus 1 and 2, and insert a plug-in that is capable of reversing polarity on one of the channels (any EQ or compressor plug-in with a polarity-reverse option will work). Insert the same plug-in onto both stereo aux tracks to avoid possible latency issues, as left and right must perfectly combine to achieve the correct combinations and phase cancellations.

On the M stereo aux input, leave both sides of the plug-in in-phase and output only to bus 3. For the S aux input polarity-reverse the right channel of the plug-in and output only to bus 4. Also reduce the level of both M and S faders by 6 dB to avoid clipping and to provide the appropriate level reduction so that the output is not doubled. The stereo file is now *encoded* into *middle* and *sides*.

DECODING (DAW EXAMPLE)

Again create two stereo aux tracks, and set the inputs for both to bus 3 and 4. The first one we will use as L and the second as R. On the L channel, keep both sides in-phase; on the right channel, reverse the phase of the right side of the plug-in. The outputs from these channels can then go directly to another stereo aux track (set inputs on bus 5/6) for further processing, or you can send the output directly to the DA converter by hard panning the L track left and R track right.

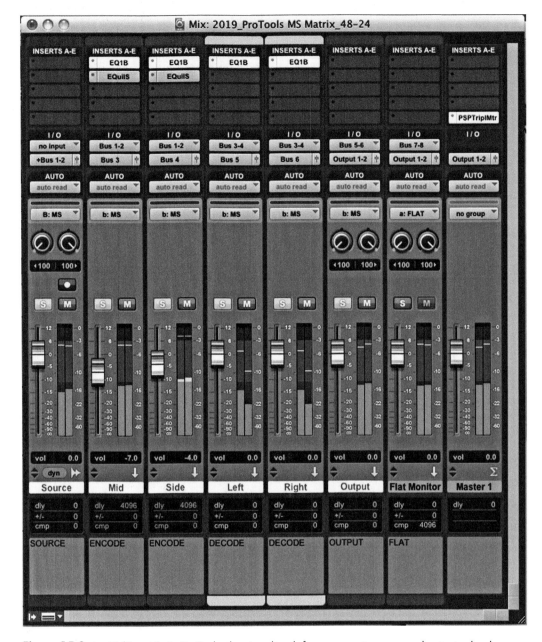

Figure 15.2 An M/S matrix in ProTools, the signal path from source to processed output soloed.

Once all of this is set up, try reducing the M and S channels to hear the effect. Reducing M should make the stereo field sound 'wider,' increasing it more 'monophonic.' Next try inserting various plug-ins in both the M and S channels. You can bypass the plug-in in the channel where you are not going to use the effect; it's just there to ensure that the latency is the same between both channels. Listen to the effect of EQing or compressing the M channel versus the S channel.

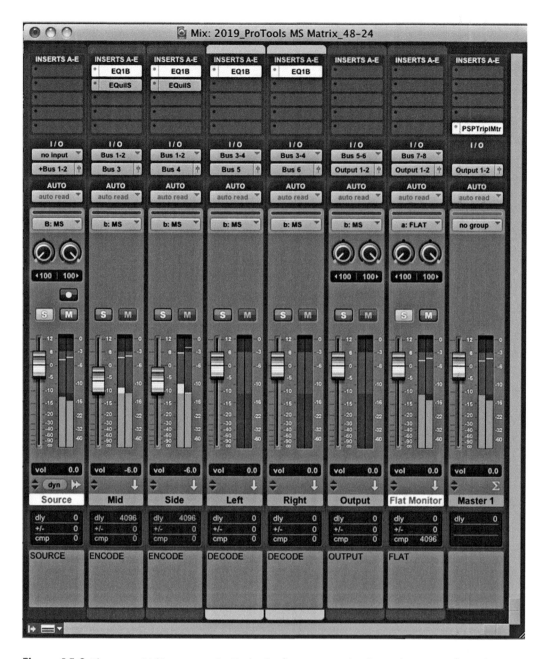

Figure 15.3 The same M/S matrix in ProTools, the flat source soloed to A/B against the M/S processing.

PRACTICAL APPLICATIONS OF M/S IN MASTERING

M/S is an extremely powerful tool, and allows for incredible flexibility in isolating and adjusting or enhancing aspects of the mix or audio image while leaving other elements largely unchanged. This is the great conundrum of audio mastering—changing

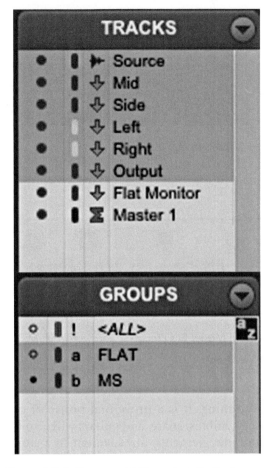

Figure 15.4 The same M/S matrix in ProTools showing the tracks and groups consisting of one audio track and five auxiliary inputs for the M/S matrix (named MS group), and also the source and flat monitor auxiliary input tracks for the FLAT group.

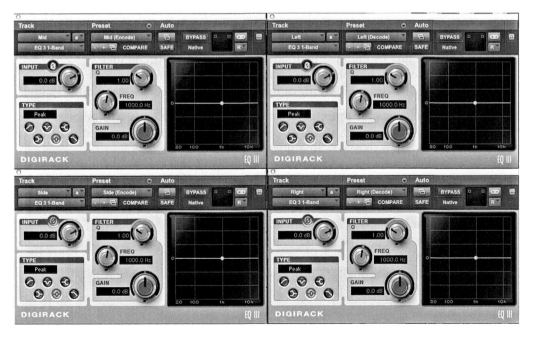

Figure 15.5 Four instances of the Digirack EQ3–1Band plug-in to set up correct phase relationships in the ProTools M/S matrix (Figures 15.2–15.4). Note that Side (encode) and Right (decode) channels are polarity-reversed via these plug-ins.

Figure 15.6 SPL Gemini Mastering M/S processer.

Source: (courtesy SPL)

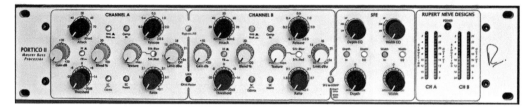

Figure 15.7 Neve Portico II Master Bus Processor hardware with M/S functionality.

Source: (courtesy Rupert Neve Designs)

anything changes everything. It is a proverbial house of cards; for example, if you brighten the vocal, the cymbals, snare, and guitars will also be affected. Or if you seek to enhance any of the mix elements, any element in a similar frequency range will be affected. Following, I list some practical applications of M/S processing you can experiment with.

1. Brighten a lead vocal without affecting the cymbals or other high-frequency stereo instruments (and vice versa)
2. Enhance guitars by boosting mid-range frequencies without affecting the vocal or snare
3. Enhance the bass without muddying the stereo guitars
4. Add body to stereo keyboards without muddying the lead vocal
5. Widen the mix by enhancing or adding level to the side channel
6. Lift out the vocal by enhancing or adding level to the mid-channel
7. Adjust the kick drum by adding or removing low frequencies in the mid-channel

There are myriad options for optimizing and adjusting a stereo mix with M/S, and the best approach is to experiment. Care is in order when utilizing M/S; remember to reference the flat mix often so that your adjustments are not excessive.

M/S HARDWARE PROCESSORS FOR MASTERING

1. t.c. electronic M6000 (see Figure 4.35)
2. Dangerous Mid-Side or Master (see Figure 3.7)
3. Maselec Mastering Consoles
4. SPL Gemini Mastering M/S processer (Figure 15.6)
5. Neve Portico II MBP (Figure 15.7)

Figure 15.8 The Waves Center Plug-in allows for a variety of frequency- and level-based M/S adjustments.

Figure 15.9 Designed by Bob Katz, the UAD K-Stereo Plug-in allows for ambience recovery, ambience EQ, and M/S balance adjustments.

Source: (courtesy Universal Audio)

Figure 15.10 The brainworx Dynamic Equalizer executes transparent M/S de-essing with extensive key adjustments, and also does a variety of dynamic EQ functions whereby the EQ settings interact with the dynamics of the music and engage as needed.

Source: (courtesy Universal Audio)

M/S PLUG-INS FOR MASTERING

1. Massey Bundle
2. Waves Center (Figure 15.8)
3. UAD K-Stereo (Figure 15.9)
4. Brainworx EQ and Dynamic Compressor (Figure 15.10)
5. DMG Equilibrium (see Figure 4.13)
6. Ozone Mastering Bundle
7. T-Racks Mastering Bundle

CONCLUSION

M/S processing embodies an *advanced mastering technique* that every Mastering Engineer must understand, gain proficiency with, and possibly configure into their *mastering system*. It may not be a regular weapon of choice or necessary for all projects, but when used judiciously, it offers impressive and impactful mastering results.

EXERCISES

1. Using an M/S plug-in or hardware setup in a mastering chain, brighten primarily the lead vocal of a mix.

2. With the same setup, widen the stereo image of the mix. Verify that it still sounds good in mono and that in stereo there isn't information emanating from beyond the speaker edges.

NOTES

1. Bob Norberg was an Engineer/Mastering Engineer at Capitol Records from 1968–2002.
2. Bobby Owsinski (2005).
3. On Nancy Wilson, *Anthology*, Capitol Records 72435-24427-2-0 (2000) and Iggy Pop, "Lust For Life," from *A Million in Prizes: The Anthology*, Virgin 09463-32920-2-4 (2005), I implemented this technique for certain tracks.
4. Tom Schlum was Chief Technician at Capitol Studios from 2006–2012.
5. EMI Archive Trust (December 16th, 2013) *Alan Blumlein and the Invention of Stereo*, Hayes, Middlesex, United Kingdom: EMI Archive Trust. See www.emiarchivetrust.org/alan-blumlein-and-the-invention-of-stereo/
6. Daniel Keller (August 9th, 2018) *Mid/Side Mic Recording Basics*, Scotts Valley, CA: Universal Audio. See www.uaudio.com/blog/mid-side-mic-recording/
7. You can also designate the left channel to be polarity-reversed, but be certain that during the decode process, the same left channel is sent to the polarity-reversed matrix channel.

CHAPTER 16

In-the-Box Considerations

An all-digital *in-the-box* (ITB) mastering setup, whereby all processing is done with plug-ins, represents a worthwhile approach to explore. With an ITB system, you can work with one DAW that can function as both the PBDAW and the RDAW. Naturally, ITB mastering requires a digital source file. Benefits of ITB are: cost-effectiveness, easy recall, and a minimized *mastering footprint* on the source audio—as neither a *DA converter* (other than for DAW monitoring) an *AD converter*, nor expensive outboard equipment, are necessary.

DRAWBACKS

Among drawbacks are that the result may lack some depth, dimension, frequency extension, and vitality. My experiments comparing the same source file mastered through a *hybrid digital/analog system* and an *in-the-box* version almost unanimously favor the *hybrid* result. I do ITB mastering a low percentage of the time, but it remains an effective option for certain projects or revision requests. ITB mastering allows the Mastering Engineer to achieve the desired target level, EQ adjustments, dynamics management, and image/width enhancement, with a result that is well correlated to the flat mix.

WHEN TO USE ITB MASTERING

Always disclose to your client a decision to approach the mastering this way, as many artists, engineers, and producers specifically seek an analog mastering chain, while others are less concerned. Situations that favor ITB mastering are a client mix on the hot side (+10dBu–+11dBu on your VU meter—likely with some peak-limiting by the mix engineer), a mix that the client loves as-is (they may have grown accustomed to the overall sound), and a mastering revision request with numerous details or concerns about the first *hybrid mastering* pass. These situations reveal that the mastering process possibly altered the *listening experience* excessively for the client's taste. This may occur because the conversion to analog and back to digital, along with the coloration from the analog chain, may alter the qualities and characteristics of the mix enough to disorient the client. If they ask for a revision, listen well to their feedback and decide upon either subtle shifts/adjustments to your original approach or an altogether alternate approach, such as ITB mastering. Experience and judgment enters into how you interpret their comments and select a plan for the revision.

BASIC AND ADVANCED SETUPS

As with an analog mastering chain, one's initial attempts at ITB mastering are best served using fundamental mastering tools (a compressor, equalizer, and brickwall limiter). I prefer to set up my PBDAW (ProTools) with a separate left and right channel (to verify left and right channel integrity through the mastering system). As a result, I open the plug-ins on the stereo master fader buss. Be aware that nearly all of the advanced mastering methods outlined in Chapter 14 are available virtually ITB. Parallel EQ is especially powerful and effective (see Figure 14.4 for an advanced ITB setup). This is due to detailed parameter adjustments available in EQ plug-ins to create frequency bands of EQ to blend in with the source audio, and one-tenth-of-a-dB fader level adjustments for the parallel blend faders. Always remember to have delay compensation engaged in ProTools with parallel processing, or you will have phase issues and smearing of the image due to delay between main and parallel audio signals.

DIGITAL HARDWARE OPTIONS

A similar approach to ITB mastering, but involving greater procurement cost, would be an outboard chain with all-digital hardware, such as devices from Weiss or t.c. electronic (see Figures 4.12 and 4.35). These are powerful multi-faceted *digital signal processing* (DSP) devices. Some mastering studios are set up with only a dedicated all-digital hardware chain, but they seem rare. I generally find that clients who may seek these high-end digital devices are also interested in high-end analog equipment for their projects.

SAMPLING THEOREM AND ITB

With ITB mastering, the DAW operates at a single *sampling frequency* and *bit depth*, usually determined by the source file resolution. If the source file is delivered at 44.1kHz or 48kHz, you can *up-sample* via *sample rate conversion* to a higher resolution (88.2kHz—24bit, 96kHz—24bit or higher) so that all the plug-in DSP is completed at the higher *sampling frequency* and *bandwidth*. Note that the bandwidth represented by the *Nyquist frequency* of the lower resolution source file will not change or increase. The advantage of *up-sampling* is that the DSP occurs at a higher *sampling frequency*, allowing for the capture of the mastered file at high-resolution specifications, even if the *Nyquist frequency* limits bandwidth.

Also note there is no way to effectively increase the bit rate of a 16bit file to 24bit in the digital domain; it must be sampled at that resolution at the original point of *AD conversion* and capture. There are conflicting theories about SRC, and concerns about the process altering a source mix (potentially adding anomalies), so some Mastering Engineers avoid it. Others see no harm and avail themselves of sophisticated SRC software such as Weiss Saracon.

Another clear advantage of an ITB system is that recalls or revision adjustments are lighting fast, as the plug-in chain can be easily saved in the DAW session. A final key point to consider is that for those compelled to develop mastering skills—ideally with access to a network of music creators to draw clients from—an effective ITB mastering can be quicker to setup and implement without the substantial costs of traditional analog mastering equipment.

Figure 16.1 A basic but very effective ITB mastering chain consisting of the: UAD Fairchild 670, DMG Equilibrium, and UAD Precision Limiter. This makes use of the Primary Colors of Mastering introduced in Chapter 1.

Source: (courtesy Universal Audio and DMG)

CONCLUSION

As elucidated previously, even if a Mastering Engineer routinely uses a hybrid analog/digital system, there is a place for ITB mastering in their arsenal of approaches. I've identified the benefits and drawbacks so that the option may be duly considered. ITB represents a starting point for developing audio mastering skills, and/or an approach for when 'plan A' does not resonate with the client. Excellent plug-ins for ITB mastering are bundles or mastering-centric offerings from UAD, Waves, T-Racks, DMG, and Voxengo.

EXERCISES

1. Set up a basic ITB mastering session using The Primary Colors of Mastering (EQ, compressor, and BWL). Take care to avoid over-levels by fine-tuning *gain structure* through the plug-in chain. Capture the result, and describe the effect and quality of the enhancements.

2. Set up an advanced ITB mastering session with one instance of parallel EQ or compression. Capture the result, and describe the effect and quality of the enhancements.

CHAPTER 17

De-Noising/Audio Restoration (Out, Damned Spot!)

The domain of a Mastering Engineer includes de-noising and audio restoration skills, making an overview important to explore. The common features of modern de-noise plug-ins (or standalone hardware or software onboard your DAW) are algorithms that will: de-buzz, de-click, de-clip, de-crackle, de-hiss, de-hum, perform spectral editing/repair, and provide tools for vinyl restoration (Figure 17.1). These tools allow the Mastering Engineer to identify, isolate, and remove unwanted noise or artifacts and restore audio. The applications of audio restoration include TV/film dialogue, cleaning up multi-track recording channels, and application of master file de-noising. In this chapter, I will explore four options for denoising and audio restoration.

SONIC SOLUTIONS NONOISE

The *Sonic Solutions NoNoise* plug-in suite (for the Macintosh platform) has a long history as an effective audio restoration option. It must be noted that the first commercially embraced de-noising system I can recall was the original Sonic Solutions NoNoise. In the mid-1990s when I began mastering at Capitol (before the proliferation of plug-ins), we were using the hardware/software version of this system. At that time, it was championed and embraced by one of our then senior Mastering Engineers, Bob Norberg. It implemented a three-step methodology which included de-hiss, de-click, and de-crackle algorithms. The de-hiss algorithm used a pure hiss 'noise estimate' (ideally from a silent area of your audio file with mostly noise and no or very little audio) to then interpolate and remove that same noise from the entire file in a background DSP process. The user could adjust the intensity of the de-noising with threshold, reduction, and other user parameters. This would achieve some dramatic results on older analog recordings besotted with tape hiss or other artifacts. They would subsequently play back with nearly crystal-clear fidelity. A similar robust and effective de-noising algorithm has evolved into the current Sonic Solutions plug-in format.

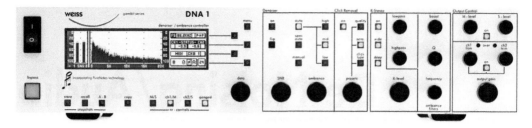

Figure 17.1 The Weiss DNA-1 is a standalone hardware de-noiser that includes a broadband de-noiser, a de-clicker, and a de-crackler. It also has the same K-Stereo ambience recovery algorithm from the UAD plug-in (see Figure 15.9). The output section consists of an additional M/S encoding/decoding matrix for final stereo width shaping.

Source: (courtesy Weiss)

Similarly, the de-click and de-crackle passes would allow for samples of the unwanted noise to be taken and then selected for the specific algorithms to interpolate and remove. Bob discovered that by setting the parameters for a gentle intensity and then running multiple passes, the best results were achieved. Invariably, there was a point where the de-noise algorithm would leave a watery or 'phasey' sound on the audio (referred to as *artifacts*) that was worse than the original. This meant that selecting the best pass was part of the process. But if the parameters were dialed in correctly, the results were often impressive.

Sonic Solutions NoNoise represented a groundbreaking de-noising system and remains a standard in professional mastering. However, as time went on, the zeitgeist shifted, and the belief grew that relevant sonic information was also being removed during the de-noising passes. Additionally, the Sonic Solutions hardware systems fell out of favor because it used a proprietary digital file system (not .wav or .aif) that could only be played back on another Sonic system. Finally, the original Sonic system was limited to standard resolution, and with the proliferation of high-resolution digital audio in the 2000s, it became passé. Sonic later re-tooled the technology to run as it does today on Macintosh-based DAWs. I can recall using the de-hiss function most often on certain introductions and fade-outs of projects from original analog tapes.

IZOTOPE RX7

Izotope RX7 is a modern DAW-hosted de-noise software plug-in that also functions as a standalone application. It has extensive audio repair, learning, and re-balancing options. It also has options to remove various sounds/noises such as 60-cycle line hum, clips, or other audio artifacts in the recording via spectral repair. Quite powerful, it functions with .wav or .aif files in all sampling frequencies and bit depths. I have found that the plug-in de-clipper is excellent at reintroducing dynamics on a peak-limited file, and it also softens the high-frequency extension, making it a viable option for creating MFiT files, files for vinyl cutting, or files for digital streaming that have a lower recommended LUFS level, such as for Spotify (Figure 17.2).

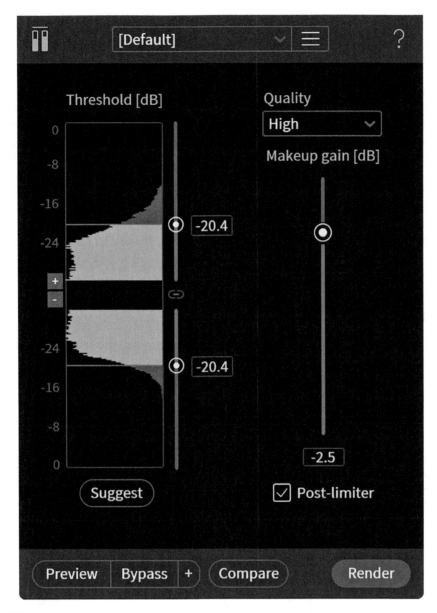

Figure 17.2 The iZotope de-clipper can eliminate inter-sample peaks, reintroduce dynamics on a peak-limited file, and soften high-frequency extension, making it a viable option for creating ADM (formerly MFiT) files.

Source: (courtesy iZotope)

CEDAR

Cedar is another company that manufactures audio restoration devices (both real-time hardware and software plug-in versions). Applications include removing unwanted artifacts (surface noises, hiss, and other analog artifacts) from digital files, as well as real-time de-noising of analog tape, vinyl album, or transcription disc during transfer to digital.

At one point in the late 1990s, I worked with a renowned engineer originally employed at Abbey Road Studios in London named Malcolm Addey. Malcolm had worked on a number of Beatles sessions in the early 1960s, and possessed an upbeat and jovial personality, along with a characteristically British self-possessed demeanor. He was accomplished and great fun to work with. We were transferring Count Basie recordings from the 1930s and 1940s from transcription discs in an era of recording when they would record a band or orchestra directly to a lathe. These recordings sounded wonderful with incredible purity and vitality, likely due to the simple microphone-to-lathe recording path and discrete electronics incorporating tubes and transformers in between. I would spray the transcription discs with Formula 409 Cleaner (a method for quiet transcription disc transfers that I gleaned from another senior engineer at Capitol, Jay Rannellucci) for the transfer, then very carefully rinse them in a large tub, and rack dry them. In addition to the Formula 409 Cleaner, Malcolm brought in a Cedar de-noise device that would remove additional hiss and pops. It literally had one threshold knob to control the intensity of the noise removal, but it worked perfectly. Cedar is still known for designing quality audio restoration hardware and software.

OPTIONS ONBOARD YOUR DAW

In DAWs designed for mastering applications, there are usually some basic de-noise functions available. I regularly use Steinberg WaveLab as my RDAW for mastering and it has a *waveform restorer* function that is excellent at removing tics, pops, and other noises that can be seen (and heard) in the waveform (Figure 17.3).

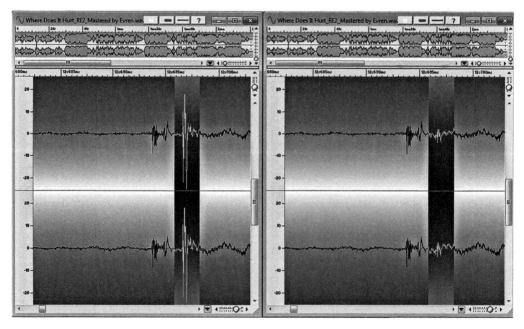

Figure 17.3 Before and after screen captures of an impulsive 'tic' in highlighted region removed by WaveLab's onboard waveform restorer algorithm.

CONCLUSION

De-noising and audio restoration experience remains relevant to a Mastering Engineer's skill set. Every mastering studio should have one of the common de-noise options available to handle any restoration issues that may arise. Common options are Izotope RX7, Accusonus ERA Bundle Pro, Antares SoundSoap+ 5, Cedar Audio Studio Complete, Sonnox Restore Collection, and Waves Audio Restoration Bundle.

EXERCISES

1. Broadband de-noise a problematic track with excessive tape hiss using Izotope RX7.

2. Using the Izotope de-clipper plug-in, reintroduce the dynamics to a loudness-maximized track (around +13dBu), creating a file that is +10dBu.

CHAPTER 18

Conclusion and Summary

There is always great excitement at a mastering session. The artist/production team is letting go of and handing off the result of an often arduous recording process to the Mastering Engineer for final enhancements. I often must politely 'shush' clients during attended mastering sessions, as they are giddy to be at the final step of their recording project. This makes for electricity in the air, but a confident Mastering Engineer must avoid distraction and access effective and streamlined approaches to successfully complete the job. The foundations and methods presented in *Major Label Mastering: Professional Mastering Process* offer exactly this benefit.

Throughout the five parts in this book, I've defined and explored: The Ten Competencies of a Mastering Engineer, Listening Experience and The Eleven Qualities of Superb Audio Fidelity; The Mastering Studio, The Three Zone Mastering System and Fundamental Mastering Tools; The Five Step Mastering Process; Macro and Micro Considerations of Mastering and Advanced Mastering Techniques. Continued practice in these areas will contribute to mastering proficiency.

Studying, grasping, and ultimately implementing the concepts and approaches I've presented in this book will be reflected in a decided improvement in the quality of the reader's mastering work. In addition, clarity and understanding will emerge regarding the assessments and procedures that occur during a mastering session. I particularly recommend steady and regular practice of The Five Step Mastering Process as outlined in Part III. Ongoing *subjective assessments* using The Eleven Qualities of Superb Audio Fidelity will serve to hone critical listening skills. Finally, implementing The Mastering Game Plan by processing and adjusting audio through The Three Zone Mastering System will engender greater confidence and effectiveness while mastering. These approaches function to help any Mastering Engineer enhance source audio more successfully.

Mastering remains a unique audio profession where technology and musical creativity intersect. The satisfaction of improving an artist's work represents a gratifying endeavor. By understanding and implementing the processes outlined herein, a Mastering Engineer will access more confidence to attempt additional approaches that sets their work apart. I hope each Mastering Engineer who studies the principles in this book experiences the exhilaration of effective audio enhancement. Happy and productive mastering!

Please email speaking or workshop inquiries to: info@majorlabelmastering.com. Additional information can be found at www.majorlabelmastering.com.

Appendix I
Common Mastering Acronyms and Terms

These are common acronyms and terms used in audio mastering that are worthwhile to learn. Many of the terms listed here are discussed in greater detail throughout the book. Some words or phrases introduce more complex concepts that may inspire further study.

ACRONYMS

AD or ADC—Analog-to-digital converter.

DA or DAC—Digital-to-analog converter.

DDP—Disc description protocol. A digital file format for a Red Book CD Master.

DFS or dBFS—Digital full scale or decibels full scale; refers to the absolute measure of a digital meter where each unit is 1 decibel (1dB).

DSP—Digital signal processing.

FFT—Fast Fourier Transform. Jean Baptiste Joseph Fourier first established that a periodic signal represented in time-domain samples can be transformed to the frequency domain as spectral coefficients. Used for creating spectrum analysis.

IC—Integrated circuit.

PCM—Pulse code modulation.

PMCD—Pre-Master Compact Disc or Plant Master Compact Disc. A compact disc, which adheres to the Red Book Standard. Contains a printed PQ sheet along with all audio and metadata for a replicating facility to create a glass master.

SPL—Sound pressure level (in dB).

VU meter—Analog meter that measures voltage level in volume units or decibels unterminated (dBu).

QC—Quality Control.

A–C

Aliasing—An undesirable phenomenon in AD converters whereby frequencies above the Nyquist frequency (half the sampling frequency) are incorrectly sampled as lower frequencies. A low-pass filter (aliasing filter) removes these frequencies.

All Bits On—A condition of a digital audio signal when there is complete usage of the digital word, often indicated by all the LED lights on a digital meter being illuminated.

Analog Signal—A continuous electrical signal whose level and frequency change in correlation with acoustic or electrical signals.

Apparent Volume—The perception of loudness to the ear, independent of level measurements. A very compressed track will have higher *apparent volume* than its VU meter and dBFS Meter readings indicate.

Artifacts—Unwanted audio byproducts of recording or signal processing, such as distortion, clicks, pops and tics.

Bandwidth—Frequency width of a band determined by two frequencies. Example: there is a 100Hz bandwidth between 100Hz and 200Hz. Important concern in equalization. A function of quality factor (Q) where BW = center frequency (Fc)/quality factor (Q).

Brickwall Limiter—A digital look-ahead dynamic control device that will not pass signal beyond a chosen dBFS level. Analog incarnations are designed with a set voltage above which signal is clipped.

Checksum—In stored or transmitted digital data, a sum of the correct digits, against which later comparisons can be made to detect errors. Used to verify the integrity of a digital file.

Class A—A common amplifier topology that uses just one output switching transistor that is continually on, making it inefficient and not well-suited for high-power applications. However, it is considered the best class of amplifier design due to excellent linearity, high-gain and low-signal distortion levels, and therefore used in high fidelity audio amplifier designs.

Compressor/Limiter Terms

Attack—Refers to the time it takes for the compressor to begin reducing the gain of a transient peak above the threshold. Indicated in milliseconds or seconds.

Peak—A peak sensing compressor responds to the instantaneous level of the input signal. While providing tighter peak control, peak sensing might yield very quick changes in gain reduction, more evident compression, or even distortion.

Ratio—The amount of gain reduction as represented by a ratio of input level above the threshold to the amount of output level above the threshold. A ratio of 4:1 means for 4dB of signal gain above the threshold, there will be 1dB of output gain.

Release—Refers to the time it takes for the compressor to release gain reduction of a transient peak and restore input level below the threshold. Indicated in milliseconds or seconds.

RMS—Acronym for root mean square. Found in amplitude measurements and compressor functions and represents an average.

Threshold—The level above which gain reduction takes place, and below which there is no gain reduction.

D–F

Decibel (dB)—One-tenth of a bel; 10 times the logarithm of a power ratio: intensity level (IL) = 10 log (P_1/P_2) dB. Where P_1/P_2 are values of acoustical or electrical power. It is a convenient measure, as it follows how humans hear the loudness of sounds.

Digital Audio Basic Terms

Bit Depth—In digital audio using pulse code modulation (PCM), it describes the number of bits of information in each sample. Pertains to dynamic range at 6dB per bit; 24bit audio has 144dB of dynamic range, 16bit audio has 96dB of dynamic range.

Clock—Refers to the crystal oscillator that sets the sampling frequency (S) of an AD converter, DAW, or any digital audio equipment. The clock of a digital playback system must coincide with the sampling frequency of a digital audio file (.wav or .aiff) for it to play back at the correct speed. If the system clock is below the file SF, the file will play back slower.

Digital Signal—A numerical representation of an analog signal using discrete time sampling and amplitude quantization.

Dither—An intentionally applied form of noise used to randomize quantization error. In a PCM digital system, the amplitude of the signal out of the digital system is limited to one of a set of fixed values (quantization). Each coded value is a discrete step: if a signal is quantized without using dither, there will be quantization distortion related to the original input signal. In order to prevent this, the signal is dithered, a process that mathematically removes the harmonics or other highly undesirable distortions entirely, and replaces it with a constant, fixed noise level. Any bit-reduction process should add dither to the waveform before the reduction is performed.

Fixed-Point—One of two arithmetic types for DSP chip design (fixed-point and floating-point). Fixed-point chips use binary integer data based in a fixed word length. 24bit and 32bit are common processor types. 24bit chips using double-precision are known as having 48bit DSP.

Jitter (Digital Clock)—In analog-to-digital and digital-to-analog conversion of signals, the sampling is normally assumed to have a fixed period—the time between every two samples is the same. If there is jitter present on the clock signal to the ADC or a DAC, the time between samples varies and instantaneous signal error arises. The error is proportional to the slew rate of the desired signal and the absolute value of the clock error. Various effects such as noise (random jitter), or spectral components (periodic jitter) can come about, depending on the pattern of the jitter in relation to the signal. In some conditions, less than a nanosecond of jitter can reduce the effective bit resolution of a converter.

Lossy/Lossless—Refers to digital file data compression. Lossy files have missing or compressed data (.mp3, .aac files). Lossless files have all their data intact (PCM file such as .wav or .aiff).

Pulse Code Modulation (PCM)—A lossless digitizing system whereby an analog waveform is sampled, quantized and coded as binary numbers representing waveform amplitudes at sample times.

Quantization—The process of measuring an analog audio event to form a numerical value. The round-off error (the difference between the actual analog value and the quantized digital value) introduced by quantization is referred to as quantization error in AD conversion.

Sampling Frequency (S)—Derives from sampling theorem, is the frequency per second at which quantization occurs. Must be at least twice the bandwidth of a sampled signal. Half the S is known as the Nyquist frequency ($N = S/2$).

Word Length—Unit of data used by a digital processor. Also refers to a fixed-size group of bits that are handled as a unit by the hardware of the processor.

32bit Floating-Point—A computer number format that occupies 4bytes (32bits) in computer memory, and implements a scaling value via a floating-point. Instead of a linear data word, it uses a non-uniform quantizer to create a data word divided into two (the mantissa and the exponent). This results in efficiency in DSP operations, more dynamic range capability, and less quantization error.

Discrete Electronics—Refers to singular electronic components not etched into the substrate of a silicon wafer known as an integrated circuit (IC). Capacitors, resistors, transistors, and inductors are examples.

Expander (downward and upward)—A downward expander performs the opposite function of a compressor, increasing the dynamic range of the audio signal. Downward expanders are generally used to make quiet sounds (valleys) even quieter by reducing the level of an audio signal (by the ratio) below the threshold. A noise gate is a type of downward expander with an extremely high ratio that will mute signal below the threshold. An upward expander also increases dynamic range, but by raising the signal level above the threshold.

Fletcher–Munson—Harvey Fletcher and Wilden A. Munson were researchers who first measured equal loudness contours in 1933. The curves are often referred to as Fletcher–Munson Equal Loudness Contours, which illustrate that the human ear is most sensitive to loudness perception between 2kHz and 5kHz, largely due to the resonance of the ear canal and the transfer function of the ossicles of the middle ear.

G–L

Glass Master—Created from a PMCD or DDP Master, usually at the plant. It is the first step in the CD replication process.

Image—The perceived sound field in a stereo listening environment.

Lisajous—The pattern on an oscilloscope, which represents a graph of a system of parametric equations. The pattern shows the phase of the stereo input signal.

M–R

Metadata—Data about data. In mastering refers to CD Text, ISRC codes, and SMPTE Time code.

Microsecond—One-10,000th of a second.

Millisecond—One-1,000th of a second.

Nyquist Frequency—Named after the Swedish-American engineer Harry Nyquist, it is half the sampling frequency of a discrete signal processing system. $N = S/2$ where N is the Nyquist frequency and S is the sampling frequency.

Operational Amplifier (op-amp)—A high-gain differential amplifier. Op-amps are a fundamental building block of analog electronic equipment.

Pink Noise—A signal simultaneously containing all frequencies from 0–20kHz. It is used to calibrate different speakers to reproduce matching sound pressure levels

(SPL). Pink noise signal is equal energy per octave. The point of pink noise is to distribute the energy according to how we hear. **Example:** The difference between 100Hz and 200Hz is one octave. The difference between 5kHz and 10kHz is also one octave. However, in terms of frequencies, the difference between 100Hz and 200Hz is only 100Hz, whereas the difference between 5,000Hz and 10,000Hz is 5,000Hz. The relative relationship is the same, but the actual difference mathematically is quite substantial. The pink noise energy between 100Hz and 200Hz is the same as between 5,000Hz and 10,000Hz.

PQ Sheet—A list containing all pertinent information regarding a CD Master. Artist, label, song titles, song lengths, and all corresponding metadata is included on a PQ Sheet.

Pumping—The generally undesirable artifact of a compressor or limiter as it attacks and releases the input signal affecting the output dynamic range. Often caused by an input signal that causes the compressor to react excessively, such as a kick drum.

Q—A number representing the ratio in Hertz of the center frequency (F_c) and bandwidth (BW). Used for equalization analysis and bandwidth representation: $Q = F_c/BW$.

Red Book Master—The CD Audio standard as defined in the Sony/Philips Red Book at 16bit—44.1kHz resolution. Specifies error detection, correction, and how data is stored.

Reference Level—An established level that is measurable at a certain amplitude on a VU meter and a digital meter (dBFS). Used to calibrate the mastering system for effective headroom.

S–Z

Saturation—Overloading of a signal causing audible distortion. May be used as a 'fattening' effect if added to or blended with the clean audio signal.

Second Order Harmonic Distortion—Nonlinearity (distortion) one octave above the fundamental signal that occurs when an amplifier's ability to cleanly reproduce audio signal is exceeded. Refers to the harmonic series that oscillates simultaneously at numerous frequencies (multiples of the fundamental). Highly controlled amounts of second order harmonic distortion is usually considered pleasing (see Chapter 14 on saturation, analog tape compression, and clipping).

SMPTE Code—Society of Motion Picture and Television Engineers time code to synchronize various audio playback machines together accurately. Originally added to film and video, it has been adapted to audio and provides an accurate time reference for editing, synchronization, and identification. Considered metadata. Common frame rates per second are 24, 25, 29.97, and 30.

SPL Meter—An instrument that measures sound pressure level in decibels.

Tape Saturation (Tape Compression)—A type of saturation (where third order harmonic distortion—one octave and a fifth above the fundamental—is apparent) that occurs when the signal input to a tape machine exceeds the capacity of the oxide particles on the tape to cleanly store and reproduce the recorded signal.

Transient Response—The response of a system to a change from equilibrium. In mastering, it refers to the initial attack of kick drums, snare drums, and other impulsive instruments or vocals. A common aim is to preserve transient response while mastering a recording.

Unity Gain—A setting on a gain device (including equalizers and compressors) whereby the input and output are the same amplitude.

Waves Dorrough Meter—The Dorrough Model 40 Audio Loudness Meter with its classic 'eyebrow' scale revolutionized audio monitoring by simultaneously showing both the traditional average program level, and peaks on a single scale. Waves Audio makes a plug-in version of this popular meter.

White Noise—White noise is equal energy per frequency. With white noise, there is substantially more energy between 5kHz and 10kHz than there is between 100Hz and 200Hz because it spans a wider range of frequencies, and they all contribute to the overall level per octave.

Zeroing—Using a signal generator set to a standard reference level (i.e. −14dBFS = 0VU = +4dBu) to separately playback mid-, high and low frequencies (1kHz, 10kHz, and 100Hz) through the mastering signal path to verify a unity gain condition exists between input and output of the mastering system. The input and the output of the mastering chain are measured with dBFS and VU meters to confirm that they match exactly.

Appendix II
Awards and Notable Projects

Among the thousands of projects I've mastered, a select few have broken through the proverbial noise floor to receive accolades or an award. In this appendix, I list some of these projects and discuss the back story that illuminates how the project was mastered and how I became affiliated.

AWARDS AND NOTABLE PROJECTS

Various Artists: "Gathering of Nations PowWow—A Spirit's Dance"

Award: Grammy® Award
Back Story:

In 2009, my friend Syd Alston from Disc Makers referred Derek Mathewes to me, as Derek's prior Mastering Engineer was no longer available. Derek, along with his family, organizes and runs the Gathering of Nations PowWow (and record label) in Albuquerque, New Mexico. They record all of the performances and dances, which are then mixed and mastered and released as an album. I was a little concerned when Derek explained the project to me—as I wasn't sure about how complex the mixing would be. Since I'd spent five years recording and mixing before focusing on mastering, I agreed to the task of mixing and mastering simultaneously. Derek and his family came through for a three-day session, where we completed approximately 20 songs.

In 2010, they came back again, and I mixed and mastered "A Spirit's Dance." Later that year, the record was nominated for a Grammy® Award in the Best Native American Album category. Come May 2011, in true Hollywood fashion, we were picked up from Capitol Records in a Hummer limousine and dropped off on the red carpet at Staples Center for the Grammys. To our delight, the album won. I am grateful to have contributed to such an important project—the additional recognition it received was well deserved.

Tupac Shakur (2PAC): "Lie to Kick It" From "R U Still Down? (Remember Me)"

Award: Recording Industry Association of America (RIAA) Certified 4x Platinum
Back Story:

In the early 1990s, I worked as a staff engineer at Paramount Recording Studios in Hollywood. One day in 1993, the office manager told me, "I put you on a session with this

Figure AII.1 Grammy® Award and Certificate for Various Artists, "Gathering of Nations: A Spirit's Dance"

rapper, Tupac. He's kind of blowin' up." I didn't think too much of it, as I'm pretty levelheaded and had worked with several name artists by that point, as well as on countless rap sessions. The day of the session (Studio A), I aligned the Studer A800 machine. Tupac came in and introduced himself. He was very polite, possessing a reserved and gentle demeanor. Then a two-man video crew showed up along with six or eight other

Figure AII.2 RIAA Certified 4x Platinum Award for 2PAC "Lie to Kick It" from "R U Still Down (Remember Me)"

rappers that included Richie Rich, Warren G, and Big Syke. I transferred a track to the 2″ tape and was playing it on the Urie 813 main monitors when they all began freestyling. The video crew was filming—so this freestyle session is easy to find online (I'm the one with the white t-shirt *not* freestyling). Then I cycled the track on loop playback as Tupac jotted down lyrics as the track played. I had set up a Neumann FET 47 through a mic pre on the SSL console and an 1176 compressor to record the vocal track. He rapped a few passes of the verses and hook, which I recorded, and that was it. This track, "Lie to Kick It," ended up on Tupac's first posthumous album on Amaru Records, "R U Still Down? (Remember Me)," selling over 4 million copies.

Mariah Carey: "Loverboy" Single

> **Award:** RIAA Certified Gold and Billboard #1 Top 100 Chart
> **Back Story:**

My good friend, excellent recording engineer and onetime roommate Michael Schlesinger, had engineered a number of Mariah Carey singles in 2000. He was working with producer Damian Young, who had strong connections at radio station Power 106 in Los Angeles. Fortunately for me, Michael insisted that I do the mastering at Capitol (there is always fierce competition amongst Mastering Engineers to work on name artist releases). One of these tracks, "Loverboy," was a radio hit. I remember requesting

Figure AII.3 RIAA Certified Gold Award for Billboard #1 single, Mariah Carey "Loverboy"

the Weiss DS-1 multiband compressor/de-esser/limiter for the session, and I ended up using it as an additional element in the mastering chain.

Poison: "Twenty Years of Rock"

Award: RIAA Certified Gold
Back Story:

Capitol catalog producer Kevin Flaherty enlisted me to remaster this collection of Poison songs from analog tapes. These recordings were produced in the era of ½" tape masters—a punchy and rich format to work from—contributing to the success of the mastering. The collection spanned the band's career, so was heavy with rock radio hits, also bolstering the potential of the release. The band performed in Studio A soon after the release of the collection and I had them sign my copy. In a classic *Spinal Tap* moment, the CD circulated to all band members and then back to C.C. Deville, who dutifully signed it a second time!

Iggy Pop: "A Million in Prizes: The Anthology"

Back Story:

Of the many projects catalog producer Kevin Flaherty and I worked on, this 2CD career retrospective of Iggy Pop's ranks among my favorites. Although it hasn't won an award,

Figure AII.4 RIAA Certified Gold Award for Poison, "The Best of Poison: 20 Years of Rock"

it has emerged as the definitive Iggy Pop collection. From a mastering perspective, the collection of recordings came out consistent and impactful despite spanning 30 years. As a fellow Michigander (Mr. Pop hails from Ann Arbor, where I attended the University of Michigan) and fan of early punk rock, as well, I had a pre-existing appreciation for the songs. So it naturally was a thrill for me to hold, and use as source material, the original mix tapes from Hansa Studios in Berlin, Germany, where the seminal albums "The Idiot" and "Lust For Life" were recorded. Both albums were produced by David Bowie, who also co-wrote many of the standout songs. On the song "Lust for Life," I used image-enhancing M/S encoding in parallel—and Mr. Flaherty later received a note of appreciation and acknowledgment of the sound of the project from Mr. Pop himself . . . which was fantastic to hear.

Pat Benatar: "Greatest Hits" and "Synchronistic Wanderings"

Award: RIAA Certified Gold
Back Story:

Both collections were compiled and remastered from original analog source tapes. Although the "Greatest Hits" collection went gold, a few years earlier the 3CD set "Synchronistic Wanderings" utilized many of the same analog source tapes, so the two make

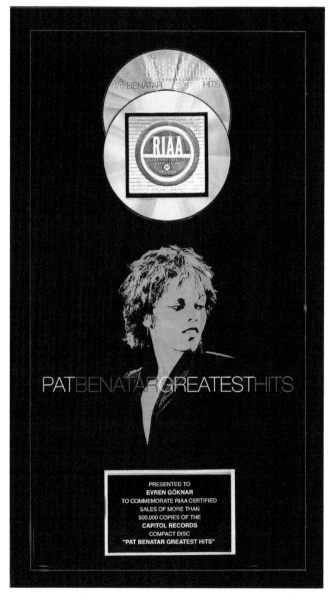

Figure AII.5 RIAA Certified Gold Award for Pat Benatar "Greatest Hits"

for an interesting comparison as they are spaced six years apart—and I mastered them in different studios at Capitol. "Synchronistic Wanderings" was mastered through the Neve mastering console in recording room 2 (RR2) and has a sound characteristic of the Neve and also a slightly lower level. "Greatest Hits" was mastered in Mastering Studio 2 using a point-to-point mastering chain without a console.

Heart: "Greatest Hits 1985–1995"

Award: RIAA Certified Gold
Story:

Figure AII.6 RIAA Certified Gold Award for Heart "Greatest Hits 1985–1995"

This was a collection for the Capitol catalog department, again utilizing original analog or digital master tapes. This was an era of songwriting collaborations for Heart, and many of these charting songs spent time in heavy radio rotation, contributing to this collection's success. The album sold steadily and achieved gold status a few years after its release.

Awards Conclusion

It's impossible to predict which project you've mastered that will go on to be highly regarded, sell millions of copies, or even win an award, so treat each project as if it is destined for wide acclaim. It is a great thrill to read in the trades that a project you mastered has received accolades. Good luck to each of you in developing musical networks and mastering compelling projects!

Appendix III
Mastering Biography

Evren Göknar was drawn to music early in life, first studying the piano and later developing his skills as a singer/songwriter and guitarist. He began his professional audio career as a recording and mixing engineer. He engineered sessions for Tupac Shakur (RIAA 4x Platinum "R U Still Down? [Remember Me]"), Tone-Loc, Mavis Staples, The Cult, Steve Vai, General Public, Carole King, and Montell Jordan (RIAA Platinum "This Is How We Do It"). Building on the skills he honed as a recording engineer, Evren shifted his focus to mastering, and in 1995, he joined Capitol Mastering.

Over the last 25 years, Evren has established himself as a top-tier Mastering Engineer, and was awarded a Grammy® Award for the "Gathering of Nations" album (Best

Native American Music Album—2011). He exclusively masters the songs for NBC's hit television show *The Voice* and also *Songland*. He has mastered projects for notable acts such as KISS, Smashing Pumpkins, Mariah Carey ("Loverboy" RIAA Gold, Billboard #1 Single), The Red Hot Chili Peppers, Lenny Kravitz, Jimmy Eat World, Iggy Pop, Heart, Megadeth, George Clinton, Eazy E, N.W.A., and George Thorogood. Among his remastering credits are seminal albums such as: Beastie Boys "Licensed To Ill" (HD), Glenn Campbell "Wichita Lineman" (HD), Heart "Dreamboat Annie" (HD), Spinal Tap (HD), Linda Ronstadt "Heart Like A Wheel" (HD), the entire Grand Funk Railroad and Queensrÿche catalogs, as well as collections by Poison (RIAA Gold) and Pat Benatar (RIAA Gold). Evren's success and experience in the music industry allow him to realize the best fidelity for a wide breadth of audio mastering projects.

In recent years, Evren has taught mastering at the university level (California State Polytechnic University Pomona) and also via independent workshops. His respect for quality music technology education and ability to present complex information in digestible segments benefits his students. These experiences informed his desire to write *Major Label Mastering: Professional Mastering Process*.

Index

Note: Page numbers in *italic* indicate a figure and page numbers in **bold** indicate a table on the corresponding page.